GRAVITY - GEOSCIENCE APPLICATIONS, INDUSTRIAL TECHNOLOGY AND QUANTUM ASPECT

Edited by **Taher Zouaghi**

Gravity- Geoscience Applications, Industrial Technology and Quantum Aspect
http://dx.doi.org/10.5772/intechopen.68576
Edited by Taher Zouaghi

Contributors
Oscar Castillo Felisola, Sultan Awad Sultan Araffa, Patrik Nemec, Hakim Saibi, Francisco Bulnes, Hongzhu Cai, Yue Zhu, Bin Xiong, Tiao-Jian Xu, Ming-Fu Tang, Guo-Hai Dong, Hassina Boukerbout, Abdeslam Abtout, Boualem Bouyahiaoui, Farida Boukercha, Khalid S. Essa, Mahmoud Elhussein, Yongqing Chen, Binbin Zhao

Published by InTech
Janeza Trdine 9, 51000 Rijeka, Croatia

Publishing Process Manager Maja Bozicevic
Technical Editor SPi Global
Cover InTech Design team

First published February, 2018
Printed in Croatia
Legal deposit, Croatia: National and University Library in Zagreb

Additional hard copies can be obtained from orders@intechopen.com

Gravity- Geoscience Applications, Industrial Technology and Quantum Aspect, Edited by Taher Zouaghi
p. cm.
Print ISBN 978-953-51-3823-5
Online ISBN 978-953-51-3824-2

For EU product safety concerns:
IN TECH d.o.o., Prolaz Marije Krucifikse Kozulić 3, 51000 Rijeka, Croatia,
info@intechopen.com or visit our website at intechopen.com.

Meet the editor

Dr. Taher Zouaghi has a Tunisian nationality and was born in the Béja Governorate. He is an associate professor and consultant in Geophysics at the Centre of Water Research and Technologies (CERTE), Borj Cédria Technopark in Tunisia. Now, he works at the Faculty of Earth Sciences, KAU, in Saudi Arabia within the technical cooperation framework between the two countries. In 2008, he received a doctorate thesis (PhD) degree in geophysics applied to structural and geodynamics, and then in 2012, he received a professorship diploma (HDR). The two diplomas were prepared at both the Faculty of Sciences of Tunis and the Borj Cédria Technopark. He conducts research into geophysics and tectonics and has published more than 45 research papers and chapters published in the international journals and conferences in the field of potential and non potential applied geophysics, forward modeling and inversion, seismostratigraphy and seismotectonics, paleoseismicity, well-logging and petroleum reservoir characterization, hydrogeophysics, salt tectonics, and geodynamics. He has supervised and still supervises many students in engineering, masters and PhD in these fields. His professional memberships include some committees and organizations (e.g., ATEIG, EAGE, AAPG, and WSEAS), and he is solicited as an expert by several scientific journals and books and scientific committees.

Contents

Preface IX

Section 1 Gravity Interpretation Tools in Geoscience 1

Chapter 1 **Gravity Data Interpretation Using Different New Algorithms: A Comparative Study 3**
Khalid S. Essa and Mahmoud Elhussein

Chapter 2 **3D Modeling and Inversion of Gravity Data in Exploration Scale 19**
Hongzhu Cai, Bin Xiong and Yue Zhu

Section 2 Gravity in Geoscience Applications 39

Chapter 3 **Microgravity and Its Applications in Geosciences 41**
Hakim Saibi

Chapter 4 **Application of BEMD in Extraction of Regional and Local Gravity Anomalies Reflecting Geological Structures Associated with Mineral Resources 73**
Yongqing Chen, Binbin Zhao, Jingnning Huang and Lina Zhang

Chapter 5 **Identification of Gravity Lineaments for Water Resources in the Crystalline Massif of Hoggar (South of Algeria) 97**
Abdeslam Abtout, Hassina Boukerbout, Boualem Bouyahiaoui and Farida Boukercha

Chapter 6 **Gravity Application for Delineating Subsurface Structures at Different Localities in Egypt 113**
Sultan Awad Sultan Araffa

Section 3 Gravity in Industrial Technology 129

Chapter 7 **Gravity in Heat Pipe Technology 131**
Patrik Nemec

Chapter 8 **Analysis of the Influence of Fish Behavior on the Hydrodynamics of Net Cage 167**
Tiao-Jian Xu, Ming-Fu Tang and Guo-Hai Dong

Section 4 Quantum Gravity 181

Chapter 9 **Beyond Einstein: A Polynomial Affine Model of Gravity 183**
Oscar Castillo-Felisola

Chapter 10 **Gravity, Curvature and Energy: Gravitational Field Intentionality to the Cohesion and Union of the Universe 203**
Francisco Bulnes

Preface

Gravity is the science of measurement and study, in space and time, of the variations of the acceleration of gravity force *g*, which is experienced by a body, at rest, on the surface of the Earth or on another planet. The basic principle of this method is essentially based on the universal attraction law formulated by Newton. Physicists currently consider Newtonian attraction as one of the interaction forms of matter, the weakest but the only one sensitive in distance. This universal force led to highly rigorous and developed mathematical theories, since, on behalf of gravitation, all matter is characterized by a single grandeur corresponding to its mass, of which the relativity theory that indicates its limits shows that we can consider it as perfectly defined. The processing of gravitation is therefore a particular discipline, which preserves the rigor of mathematical developments, without depending on the contingent and always imperfectly known properties of matter.The average value of gravity (gravity force) on the surface of the Earth, at the latitude 45°, is g = 9, 81 m/s^2 = 981 Gals.This gravitational field reflects a superposition of four fields: the Newtonian field that corresponds to the gravitational force caused by the Earth, the centrifugal force due to the rotation of the Earth or the planet and its geometrical form, the field due to the other stars (Moon, Sun, and planets), and cohesive forces. The study of gravity allows us to measure the mass distribution and consequently to deduce the internal structure of the planets.

The two first sections of this book deal with subjects concerning geoscience domains (gravity interpretation tools and gravity applications). Gravity is of utmost importance for many fields such as earth and environmental sciences, geophysics, geodesy, metrology, or particle physics. Progress in these areas is therefore closely linked to the evolution of gravity. Recently, new technological advances have improved significantly the measurement of g. In the domain of geophysics, gravity belongs to potential methods, which tends to measure the density variations. It allows the recognition of deep-seated structures by characterizing possible mass heterogeneities. Application of a certain number of corrections to the raw values leads to suppress the effect of other bodies and to bring out only anomalies due to subsoil density variations. Microgravity corresponds to measurement of very small variations in gravitational attraction of about 5 to 100 μGal (1000000 Gal or 10^8 m/s^2), in the superficial layers. It is one of the most used methods for detection of small bodies or voids. It is indeed sensitive to contrast density, geometry, and extension of the masses.The main difficulties in implementing the method stem from its sensitivity to many undesirable disturbing effects and the non-uniqueness of interpretations.

Gravity-based geophysical studies are becoming more important because they contribute to a better understanding of the geological sciences in general, particularly in subsurface exploration.The advances of ideas and concepts have contributed to the development of geophysical methods and techniques, which have led to significant progress in the study, evaluation, and valorization of deep-seated structures. Gravity anomalies made after the measurements and the compilation of the data during the survey reveal a variation of the potential field of the earth. The obtained anomalies correspond to the sum of the effects of all bodies whatever the orientation, the nature, and the densities. It is therefore not always easy to make a simple link between the anomalies and their causative structures. The use of Fourier transforms (mathematical operators) has become a common practice in geophysics, and their utility in gravity has been demonstrated since the 1960s. Indeed, gravity field can be represented exactly by a development in Fourier series, used to calculate the different transformations applied to the gravity data. Quantitative interpretation of gravity survey data corresponds to rigorous analysis, which is carried out on a routine basis only when simple geometric models are utilized to represent the subsurface sources. The quantitative interpretation is often carried out by forward modeling and inversion of causative sources that could theoretically reproduce the anomalies recorded in the survey. However, modeling and inversion are not easy to be operated on, particularly because, for a given set of geophysical data, an infinite number of distributions of the subsoil physical properties may be appropriate. The geophysicist is then recommended to choose the inversion parameters and to constrain his model in the most realistic way possible in order to obtain the most probable model.

The third section of the book deals with the aspect of industrial technology; two studies are presented: the first concerns the heat pipe and the second concerns the fish behavior in cage. Gravity is one of the phenomena, which affects the heat transport capacity of the heat pipe. Heat pipe technology is one form of technologies which is more and more used in technical application. It is mainly used in the area of heat transfer such as electronic cooling, heat recovery from technical processes, and heat recovery from ground and sun to use in heating system. Heat pipe experiments based on gravity are presented and discussed. The chapter describes the principle, design, and construction of the gravity heat pipe (GHP) and the wick heat pipe (WHP). It is an experimental and mathematical study of the influence of gravity on the thermal transport capacity of wick heat pipes with the change of working position. Another chapter concerns the comportment and distribution of fish at night and during the daytime. The study allowed to propose a net-fluid interaction model, showing the effects of the fish behavior on the flow field around the net cage and deformation of the net cage. This type of study allows the development of a technico-economic reference frame for marine farming activities, in order to establish a list of measures ensuring the maintenance and conservation of the species.

The book ends with a section that deals with quantum gravity, a branch of theoretical physics attempting to unify quantum mechanics and general relativity. The fundamental physics knew two revolutions during the twentieth century, with the introduction of the theories of general relativity and quantum mechanics. Quantum gravity is defined as the hypothetical unification of these theories. It is customary also to present this problem as that of the unification of the four forces: electromagnetic, weak nuclear, strong nuclear, and gravitational. Another point of view is to see quantum gravity as the theory of the unification of the three fundamental constants. Two chapters are presented: in the first chapter, the author shows the effective field equations for a recently formulated polynomial affine model of gravity, in the sector of a torsion-free connection, and accepts general Einstein manifolds with or without cosmological constant as solutions. The author presents a short review of the polynomial affine gravity, whose field equations (in the torsion-free sector) generalize those of the standard general relativity. In the mentioned approximation, the field equations coincide with (part of) those of a gravitational Yang-Mills theory of gravity, known in the literature as the SKY (Stephenson-Kilmister-Yang) model. The second chapter discusses the quantum operators as which are diffeomorphisms of gravity to create complex bodies in the scaled of the conforming of the fragmented Universe. The complex Riemannian structure, the space-time model, and the K-invariant G-structure of the orbits, are obtained as consequences of the diffeomorphisms. These later are considered as field equations to the energy-matter tensor density for each case of the gravitational field.

This book is particularly valuable insofar it presents multidisciplinary approaches. The authors provide basic methods of gravity, which are very interesting in geoscience and industrial technology fields. Both theoretical and computational quantum gravity, new techniques are also presented through examples of applications and numerous scenarios. The chapters show the importance and advantages and sometimes the shortcomings of the different techniques used.

This precious product could not be finalized without the participation and interest of many collaborators. I would like to acknowledge the authors, not only for their contribution but also for many discussions on different aspects of the book's content. I would like to appreciate their enthusiasm for putting their knowledge and experience at the service of such an arduous but necessary task. I would like to express our gratitude to Ms. Maja Bozicevic, Publishing Process Manager, and Ms. Ana Pantar, Senior Commissioning Editor, from InTechOpen Book, for their highly valuable support and help in the many phases of the book preparation and edition processing.

Taher Zouaghi
Associate Professor and Consultant in Geophysics
CERTE, Technologic Pole of Borj Cédria, Soliman, Tunisia
Faculty of Earth Sciences, King Abdulaziz University, Jeddah, Saudi Arabia

Section 1

Gravity Interpretation Tools in Geoscience

Chapter 1

Gravity Data Interpretation Using Different New Algorithms: A Comparative Study

Khalid S. Essa and Mahmoud Elhussein

Additional information is available at the end of the chapter

http://dx.doi.org/10.5772/intechopen.71086

Abstract

Gravity data interpretation is useful in exploring regions that have different geological structures, which contain minerals, ores and oil deposits. There are different numerical methods for the model parameters (depth (z), origin location (x_o), shape parameter (q) and amplitude coefficient (A)) evaluation of a covered structure such as gradient method, particle swarm optimization technique and Werner deconvolution method. In this study, application of these methods is utilized to appraise the model parametric quantity of the covered structures. The application of these methods was demonstrated by different engineered data without and with various range of noise (5%, 10%) and applied for a real example from Egypt. The result values of each method were compared together and with those published and drilling information.

Keywords: gravity anomaly, depth, werner deconvolution, PSO, gradient method

1. Introduction

Gravity method is a non-ruinous geophysical procedure that measures contrasts in the gravitational field of the earth at many various areas. It has much beneficial utilization in hydrocarbon exploration, mineral prospecting, archeological investigations, environmental applications and crustal imaging [1–11]. The main objective of the gravity interpretation is evaluating the model parameters (depth, amplitude coefficient, origin location, and shape parameter) of gravity oddities delivered by basic geometrical formed structures (spheres, cylinders). Clarification of gravity data is constantly connected with the ill-posed and non-unique problems. To overcome these issues, we find a preferred geometry to subsurface structures with a known density followed by the inversion processes [12, 13]. Understanding of gravity data can be performed utilizing basic geometrical models, forward modeling and inversion.

Analytical formula for basic geometrical shapes and many approaches have been produced to translate the gravity anomaly expecting the body of basic geometry (sphere, horizontal cylinder and vertical cylinder). These techniques have varying complexity in the interpretation.

All different simple models may not be found in real subsurface geological situations, they usually are preferred in practical inversion of many isolated sources. The target of an inversion process is to recover the converse parameters of the model (depth, amplitude coefficient, origin location and shape factor). Many scientists showed and discussed several graphical and numerical approaches developed in past and significantly in the present time [10, 11, 14–32]. However, the disadvantages of these methods that depend on characteristic points and curves subject to person errors in calculating the inverted parameters of the subsurface structures which can prompt significant errors in assessing the inverse parameters of the covered structure [10, 11]. Thus, the outcomes from these techniques need the accessibility of density information as a noteworthy aspect of the commitment, alongside similar depth information got from geology and/or geophysics. Consequently, the resultant model can shift comprehensively relying upon these factors since the inverse problems are not well-postured and are along these lines unsteady and non-unique [33].

The interpretation of the gravity data is attempted here using three methods: the gradient method [34], the particle swarm optimization and Werner deconvolution method [21]. Analysis of the gravity anomalies can allow obtaining more detailed information on the geological structures that partially outcrops or covered totally in depth. In overall, these different methods are utilized in this work to searching the sources nature of gravity anomalies. The results of applied three different methods are compared together. A synthetic example without and with various level of noise (5% and 10%) used to show the stability of these methods. The proposed techniques are additionally tested on a gravity data from Egypt. To judge satisfaction and fulfillment of these approaches is finished by contrasting the acquired results with other accessible geological or geophysical information in the published literatures.

2. The methods

Different three algorithms used to interpret the gravity anomaly (mGal) produced by most common three shapes (spheres, horizontal cylinders and vertical cylinders) (**Figure 1**) represented by:

$$g(x_i, z, q) = \frac{A}{\left[(x_i - x_o)^2 + z^2\right]^q}, \quad i = 1, 2, 3 \ldots\ldots\ldots N \tag{1}$$

where

$$A = \begin{cases} \frac{4}{3}\pi G\sigma z R^3 \\ 2\pi G\sigma z R^2, \\ \pi G\sigma R^2 \end{cases} \quad q = \begin{cases} 1.5 & \textit{for a sphere} \\ 1 & \textit{for a horizontal cylinder.} \\ 0.5 & \textit{for a vertical cylinder} \end{cases}$$

In the above equation, z is the depth (m), A is the amplitude coefficient (mGal × m^{2q}) that depends on the shape parameter, q is the parameter related to the shape of the body (dimensionless), x_i is the position coordinate (m), x_o is the origin location (m), σ is the density contrast between the target and the surroundings, G is the gravitational constant parameter which equal 6.67 × 10^{-11} SI units, and R is the radius of the covered body (m), as follow:

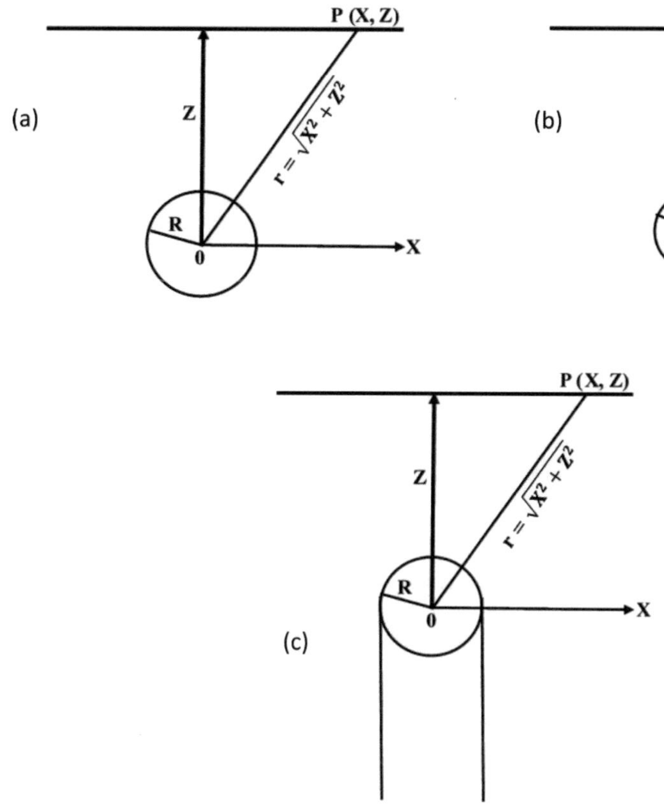

Figure 1. Sketch diagram for different simple geometrical structures: (a) sphere model, (b) horizontal cylinder model and (c) vertical cylinder model.

2.1. The gradient method

The gradient algorithm [34] depends on the utilizing the numerical fourth horizontal gradient registered from the measured gravity anomaly utilizing filter of successive window lengths to evaluate the depth and shape of covered structures. The numerical fourth gradient gravity value at point x_i is figured from measured gravity data $g(x_i)$ by:

$$\Delta g_{xxxx}(x_i) = \frac{\{\Delta g(x_i + 4s) - 4\Delta g(x_i + 2s) + 6\Delta g(x_i) - 4\Delta g(x_i - 2s) + \Delta g(x_i - 4s)\}}{16s^4}, \quad (2)$$

where s is a window length or graticule spacing.

Also, the depth computed using the following form derived from the above equation:

$$F = \frac{A}{16\,s^4}\left(\frac{1}{\left(\left(x_o+4s\right)^2+z^2\right)^q} - \frac{4}{\left(\left(x_o+2s\right)^2+z^2\right)^q} + \frac{6}{(x_o^{\,2}+z^2)^q} - \frac{4}{\left(\left(x_o-2s\right)^2+z^2\right)^q} + \frac{1}{\left(\left(x_o-4s\right)^2+z^2\right)^q}\right) = 0. \tag{3}$$

2.2. The particle swarm optimization (PSO)

PSO-algorithm was created by [35]. It's relying upon the reenactment of the apparent conducts of birds, fishes and insects in food searching. PSO-algorithm is applied in many issues, like model construction [36], biomedical images [37], electromagnetic optimizations [38] and hydrological problems [39]. In this calculation, the birds representing the particles or models, every molecule has a location vector which speak to the parameters esteem and a velocity vector. So, for a four-dimensional improvement issue, each molecule or individual will have a location in four-dimensional spaces which speak to a solution [40]. Each molecule changes its location at every movement of the operation of the algorithm, this location refreshed amid the iteration procedure considering the best location reached by the molecule which is called the T_{best} model and the best location obtained by any particle in the community called the J_{best} model, this refreshment is clarified in Eqs. (4) and (5) [41]

$$V_i^{k+1} = c_3 V_i^k + c_1 rand()(T_{best} - P_i^{k+1}) + c_2 rand()\left[\left(J_{best} - P_i^{k+1}\right) P_i^{k+1}\right] = P_i^k + V_i^{k+1}, \tag{4}$$

$$x_i^{k+1} = x_i^k + v_i^{k+1}, \tag{5}$$

where v_i^k is the speed of the molecule i at the kth cycle, P_i^k is the current I modeling at the kth cycle, rand() is an arbitrary number in the vicinity of 0 and 1, c_1 and c_2 are positive constant numbers which ascendency the person and the sociable behavior, they are typically taken as 2 [41] yet some recent researches give that picking c_1 more prominent than c_2 however $c_1 + c_2 \leq 4$ may give better outcomes [42], c_3 is the inertial coefficient which control the velocity of the molecule, since the substantial esteems may shuffle the molecules to miss up the great arrangements and the small esteems may bring about insufficient place for exploration [41], it's usually taken less than 1, x_i^k is the positioning of the molecule i at the kth cycle.

The four model parameters (z, A, x_o, and q) can be evaluated by using the PSO-algorithm to reach the misfit by using the following objective function:

$$Q = \frac{2\sum_{i=1}^{N}|T_i^m - T_i^c|}{\sum_{i=1}^{N}|T_i^m - T_i^c| + \sum_{i=1}^{N}|T_i^m + T_i^c|}, \tag{6}$$

where N is the number of data points, T_i^m is the observed gravity anomaly, T_i^c is the evaluated gravity anomaly.

2.3. Werner deconvolution method

Werner deconvolution method [21, 43] was also originally developed for magnetic interpretation. Also, Werner deconvolution has been used for gravity interpretation. The method is particularly useful when the profile anomaly of interest can be expressed as a rational function of the form of Eq. (1). As identified by [43], Eq. (1) can be rewritten in linear form as follow:

$$[g(x_i)]^{e_1}(x_i - e_4)^2 + [g(x_i)]^{e_1} e_2 - e_3 = 0, \tag{7}$$

where

$$e_1 = \frac{1}{q}, \; e_2 = z^2, \; e_3 = A^{1/q}, \; e_4 = x_o.$$

Eq. (7) is linear form in the four variables e_1, e_2, e_3 and e_4, so that a numerically remarkable arrangement can be found for them from evaluating the equation at four points.

The Root Mean Square error (RMS) between the data and model responses is evaluated as follows

$$RMS = \sqrt{\frac{\sum_{i=1}^{N}[T_i^m(x_i) - T_i^c(x_i)]^2}{N}}. \tag{8}$$

This is considered as a rule in evaluating the best-fitted model parameters (z, A, x_o, q) of the covered structure.

3. Synthetic example

Noisy-free gravity anomaly for a horizontal cylinder with A = 400 mGal m^2, z = 5 m, q = 1, m = 1 and profile length of 120 m. Our analysis begins by applying the fourth horizontal gradient separation technique (Eq. (2)) to the gravity anomaly utilizing distinctive s-values (s = 2, 3, 4 and 5 m) (**Figure 2**). By applying this inversion technique, we evaluated z and A values at different q for every s-value and after that ascertained the average depth and RMS (**Table 1**). **Table 1** exhibits the estimation consequences of the interpretation of noise free data. The assessed parameters from the proposed technique are in a decent concurrence with the model of the horizontal cylinder where z = 5 m, A = 400 mGal m^2 and q = 1. At long last, we can watch that the minimum RMS (RMS = 0 m) occurs at the true model parameters.

Because of the real data are tainted with random noise, random noise of 5 and 10% imposed on the gravity anomaly to see the effect of these noises on the inversion method. The fourth horizontal gradients were evaluated using the same s-values mentioned above (**Figures 3** and **4**). **Table 1** also demonstrates the computational outcomes of the interpretation of noisy gravity data. The average depth of 5 m and the solution with minimum RMS (0.65 mGal) gives in case of 5% noise and depth 4.9 m and RMS of 4.4 mGal in case of 10% noise. This shows that this method is useful when applied to noisy gravity data. In addition, we use

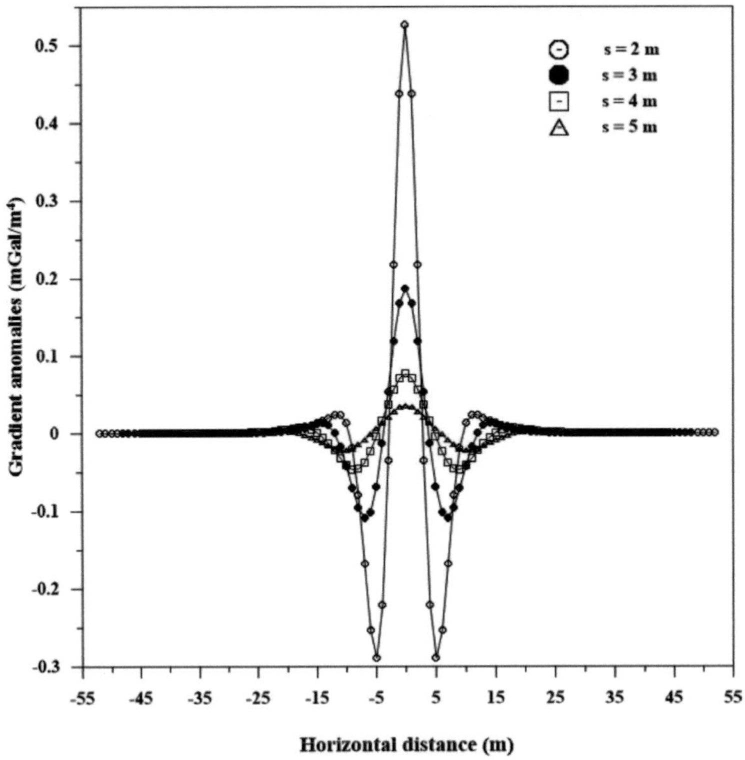

Figure 2. Data analysis of the horizontal cylinder model using the gradient method.

s (m)	Vertical cylinder model, q = 0.5		Horizontal cylinder model, q = 1		Sphere model, q = 1.5	
	z (m)	A (mGal m)	z (m)	A (mGal m^2)	z (m)	A (mGal m^3)
2	3.9	465.8	5	400	4.0	2753.37
3	3.8	461.2	5	400	3.9	3447.5
4	4.1	447.2	5	400	3.9	3916.2
5	4.2	433.1	5	400	4.2	4230.3
Average	4.0	451.9	5	400	4.0	3586.8
RMS (mGal)	17.57		0		25.71	
With 5% random noise						
2	4.2	465.8	5.1	400.0	4.1	2753.3
3	3.7	461.2	4.9	410.5	3.8	3447.5
4	3.8	447.2	5.2	417.7	3.7	3916.2
5	3.9	433.1	4.9	424.6	4.3	4230.2

s (m)	Vertical cylinder model, q = 0.5		Horizontal cylinder model, q = 1		Sphere model, q = 1.5	
	z (m)	A (mGal m)	z (m)	A (mGal m²)	z (m)	A (mGal m³)
Average	3.9	451.8	5.0	413.2	3.9	3586.8
RMS (mGal)	17.87		0.65		27.23	
With 10% random noise						
2	4.0	351.7	5.0	479.4	3.9	2079.0
3	4.0	410.5	4.8	481.1	4.0	3068.7
4	4.0	406.3	4.7	432.9	3.9	3557.9
5	3.9	404.2	5.1	422.83	4.0	3947.8
Average	3.9	393.2	4.9	479.4	3.9	3163.4
RMS (mGal)	13.65		4.4		22.29	

Table 1. Numerical results for a gravity model due to horizontal cylinder without and with two levels of 5% and 10% of random noise (A = 400 mGal m², z = 5 m, q = 1, and profile length = 120 m) using the gradient method.

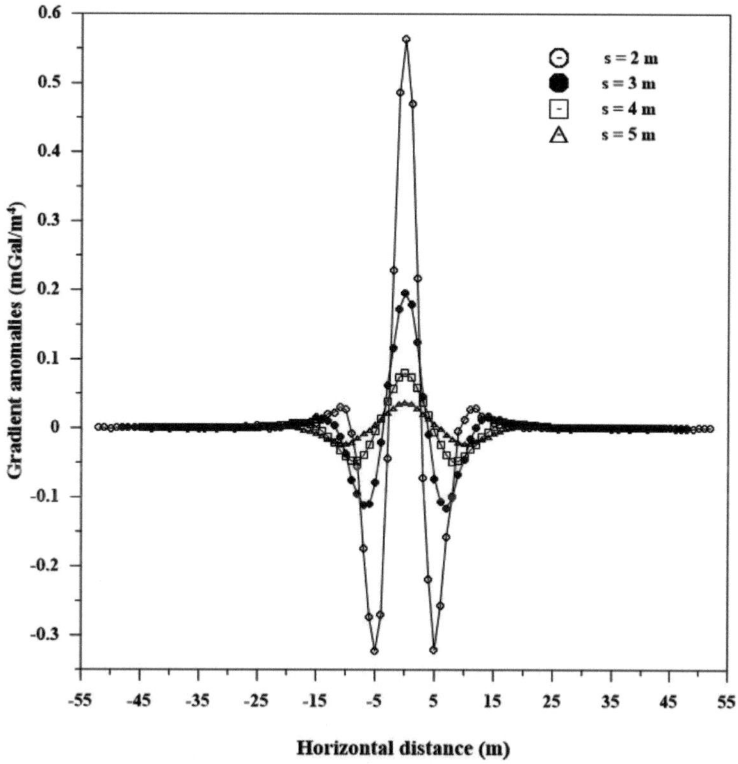

Figure 3. Data analysis of the horizontal cylinder model using the gradient method when the data contain 5% random errors.

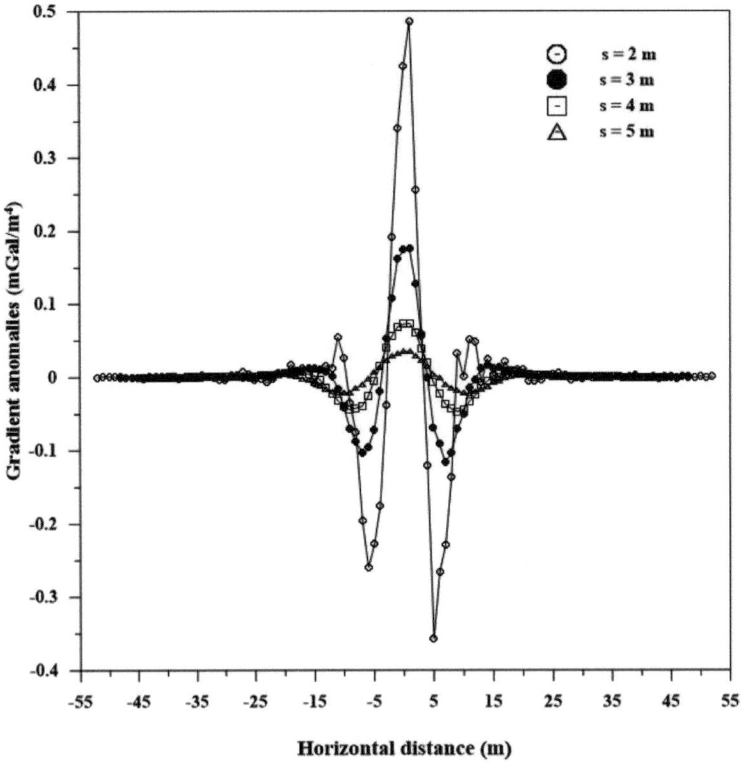

Figure 4. Data analysis of the horizontal cylinder model using the gradient method when the data contain 10% random errors.

Werner deconvolution method to the same gravity anomaly utilizing the same window size every 2 m. we used 11 clustered solutions to calculate the average estimated depth is 5 m, A = 400 mGal m^2, and q = 1 with RMS = 0 mGal. Also, as mentioned above we use the same Werner deconvolution method for the noisy gravity anomalies. The average estimated depth of the cluster solutions is 5.3 m, A = 410.1 mGal m^2 and q = 1 with RMS = 0.82 mGal in case of adding 5% random noise. Also, the average estimated depth of the cluster solutions is 5.6 m, A = 425.3 mGal m^2 and q = 1 with RMS = 1.20 mGal in case of adding 10% random noise (**Table 2**).

The PSO-algorithm was connected to the same synthetic gravity anomaly. In this circumstance, it is noise free data, so we start testing our technique using 100 models. The best model came after 700 cycles, the used extent of the parameters are showed up in **Table 3**. The assessed model parameters which control the body measurements are in good correlation with the proposed values (**Table 3**) corresponding to zero RMS. Since, the uproarious data considered as a basic part in geophysics, thusly, we applied our method to 5% arbitrary random noise gravity data caused by horizontal cylinder model appear with a particular true objective to inquire about the effect of noise corrupted data. The assessed indicate parameters (z, A, x_o, q) are presented in **Table 3. Table 3** exhibits that the RMS error is 0.32 mGal. Plus, we forced 10% of subjective random noise on the comparable synthetic anomaly. Also, **Table 3** demonstrates the inverted parameters and shows that the RMS error is 0.64 mGal.

	Vertical cylinder model, q = 0.5			Horizontal cylinder model, q = 1			Sphere model, q = 1.5		
	z (m)	A (mGal m)	x_o (m)	z (m)	A (mGal m^2)	x_o (m)	z (m)	A (mGal m^3)	x_o (m)
Average	4	458.37	0	5	400	0	3	3875.3	0
RMS (mGal)	18.04			0			55.96		
With 5% random noise									
Average	4.3	465.51	0	5.3	410.1	0	3.4	3884.4	0
RMS (mGal)	17.60			0.82			42.82		
With 10% random noise									
Average	4.5	468.26	0	5.6	425.3	0	3.6	3910.1	0
RMS (mGal)	17.28			1.20			38.25		

Table 2. Numerical results for a gravity model due to horizontal cylinder without and with two levels of 5% and 10% of random noise (A = 400 mGal m^2, z = 5 m, q = 1, and profile length = 120 m) using Werner deconvolution method.

Type of body	Parameters	Used ranges	Result	RMS (mGal)
	Without random Gaussian noise			
	A (mGal m^2)	100–700	400	0
	z (m)	2–12	5	
	q	0–3	1	
	x_o (m)	−20 to 50	0	
	With 5% random Gaussian noise			
	A (mGal m^2)	100–700	395	0.32
Horizontal cylinder model	z (m)	2–12	5	
	q	0–3	1	
	x_o (m)	−20 to 50	−0.01	
	With 10% random Gaussian noise			
	A (mGal m^2)	100–700	411	0.64
	z (m)	2–12	4.9	
	q	0–3	1	
	x_o (m)	−20 to 50	0.02	

Table 3. Numerical results for a gravity model due to a horizontal cylinder without and with two levels of 5% and 10% of random noise (A = 400 mGal m^2, z = 5 m, q = 1, x_o = 0 m and profile length = 120 m) using the PSO-technique.

4. Field example

So as to inspect the pertinence and effectiveness of the three showed methods on the real data, we have connected the three techniques to a gravity anomaly profile of Abu Roash dome area, the Northern Western Desert, Egypt (**Figure 5**). The Bouguer gravity map is situated in the West of Cairo ([44]; his Figure 11) and was mapped in 1980 by the Egyptian General Petroleum Corporation (EGPC) utilizing a density of 2.3 g cm^{-3}. The structure information is accessible from the surface geology and drilled hole data [45]. From the geology information of the area, we observe that the basement rocks (with greater prominent thickness than the above sedimentary layers) are elevated because of the high pressure in the SW direction [45]. At the Abu Roash dome, there are exposures of Cenomanian clastics at its core took after by Turonian and Senonian strata. This Cretaceous succession separated from the above Eocene sediments by an angular unconformity [45–47]. **Figure 5** shows the Bouguer anomaly profile which are opposite to the heading of compression striking NW–SE, this profile was digitized at an interim of 300 m. The Bouguer anomaly accordingly acquired has been subjected to the three various methods (the fourth horizontal gradient method, Werner deconvolution method, and the PSO-technique).

Firstly, we used the fourth horizontal gradient method to four progressive windows (s = 600, 900, 1200 and 1500 m) to obtain the inverted model parameters. The four fourth horizontal gradient anomaly profiles were gotten (**Figure 6**). **Table 4** summarized the results obtained from this method. Secondly, by applying Werner deconvolution method to the same observed gravity data, the outcomes are summarized in **Table 5**. Thirdly, a PSO-algorithm utilized to assess the interpretive model parameters of gravity anomaly profile. **Table 6** displays the ranges and results of the evaluated parameters.

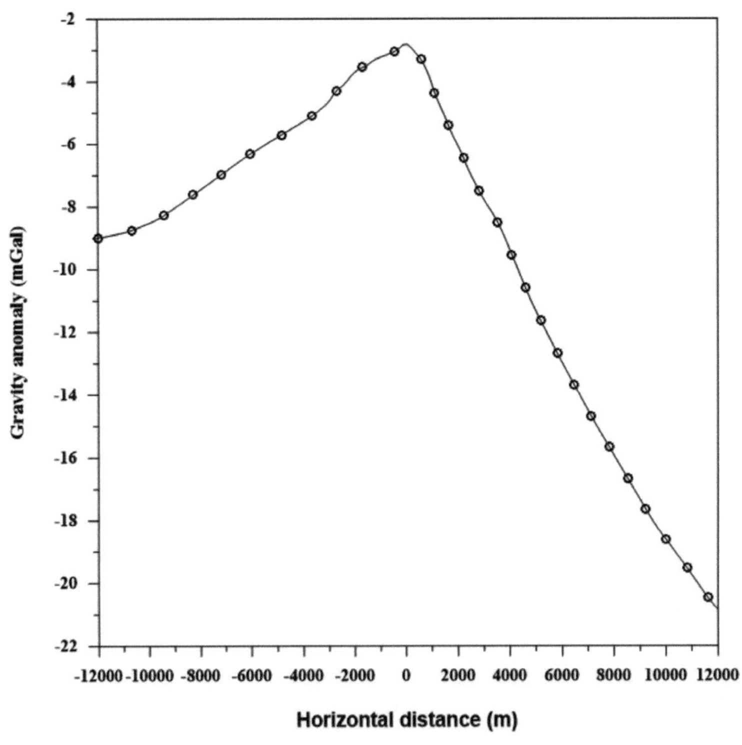

Figure 5. Observed gravity anomaly profile of Abu Roash field example, Egypt.

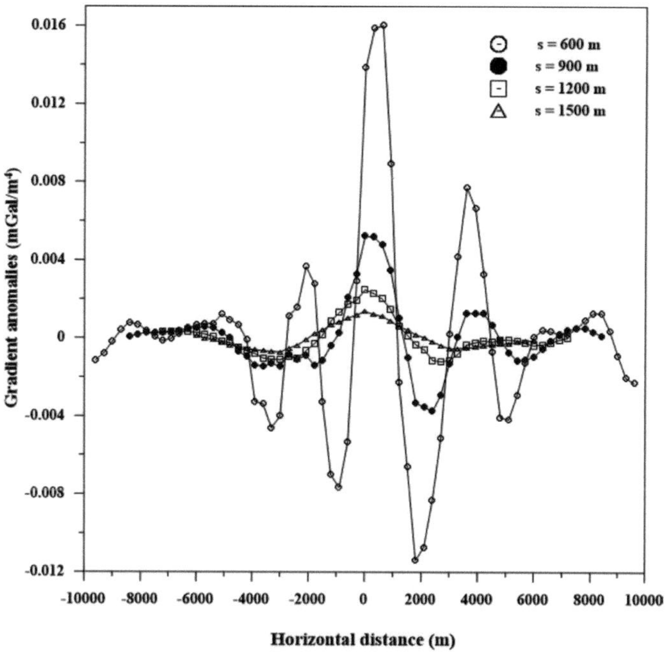

Figure 6. Data analysis of the Abu Roash field example using the present gradient method.

s (m)	Vertical cylinder model, q = 0.5		Horizontal cylinder model, q = 1.0		Sphere model, q = 1.5	
	z (m)	A (mGal m)	z (m)	A (mGal m²)	z (m)	A (mGal m³)
600	1870	−5236	2200	−6160	2520	−13,552,000
900	1890	−5292	2250	−6300	2600	−14,175,000
1200	1920	−5376	2510	−7028	2830	−17,640,280
1500	2050	−5740	2630	−7364	3290	−19,367,320
Average	1932.5	−5411	2397.5	−6713	2810	−16,183,650
RMS (mGal)	10.04		10.38		224	

Table 4. Numerical results of Abu Roash dome field example using the gradient method.

	Vertical cylinder model, q = 0.5			Horizontal cylinder model, q = 1.0			Sphere model, q = 1.5		
	z (m)	A (mGal m)	x_o (m)	z (m)	A (mGal m²)	x_o (m)	z (m)	A (mGal m³)	x_o (m)
Average	1870	−5230	0	2320	−6503	0	2750	−17,253,450	0
RMS (mGal)		10.06			10.39			246.99	

Table 5. Numerical results of Abu Roash dome field example using Werner deconvolution method.

Parameters	Used ranges	Result	RMS (mGal)
A (mGal m)	−2000 to −6000	−5200	0.17
z (m)	500–2500	1860	
q	0–2.5	0.45	
x_o (m)	−100 to 100	0	

Table 6. Numerical results of Abu Roash dome field example using the PSO-technique.

Parameters	Method		Present methods		
	[45]	[48]	Fourth gradient	Werner	PSO
A (mGal m)	1900	1620	−5411	−5230	−5200
z (m)	–	–	1932.5	1870	1860
q (dimensionless)	–	0.5	0.5	0.5	0.45
x_o (m)	–	–	–	0	0

Table 7. Comparison between the present three used method and different methods for Abu Roash field example, Egypt.

Finally, the three inversion techniques give a full picture of the model parameters instead of various techniques which did not give a totally elucidation. The results are outlined in **Table 7**.

5. Conclusions

In this chapter, three various methods were used for modeling gravity anomaly due to simple geometrical shaped. The viability of the proposed methods (the gradient method, particle swarm optimization method and Werner deconvolution method) is used on a synthetic example including noisy-free data, contaminated data with various level of noise (5 and 10%), and a real field data from Egypt. The three approaches can enhance the quality solution and convergence traits and computational adequacy. The examination of the results with drilling information and published information detailed in the literature demonstrated the prevalence of the three methods and its potential for dealing gravity issue. Later on work, we will attempt to suggest some enhanced variant of these methods to deal with issue.

Acknowledgements

We would like to thank Prof. Taher Zouaghi, the Editor, for his keen interest, imperative comments on the Article, and changes to this work. Also, we might want to express gratitude to Ms. Maja Bozicevic, Publishing Process Manager, for her help and collaboration in this issue.

Author details

Khalid S. Essa* and Mahmoud Elhussein

*Address all correspondence to: khalid_sa_essa@yahoo.com

Geophysics Department, Faculty of Science, Cairo University, Giza, Egypt

References

[1] Grant FS, West GF. Interpretation Theory in Applied Geophysics. New York: McGraw-Hill Book Co.; 1965

[2] Roy A. The method of continuation in mining geophysical interpretation. Geoexploration. 1966;**4**:65-83

[3] Nettleton LL. Gravity and Magnetics in Oil Prospecting. New York: McGraw-Hill Book Co.; 1976

[4] Ateya IL, Takemoto S. Gravity inversion modeling across a 2-D dike-like structure—A case study. Earth, Planets and Space. 2002;**54**:791-796

[5] Fedi M. DEXP: A fast method to determine the depth and the structural index of potential fields sources. Geophysics. 2007;**72**(1):I1-I11

[6] Lafehr TR, Nabighian MN. Fundamentals of Gravity Exploration. Tulsa, OK: Society of Exploration Geophysicists; 2012. p. 211

[7] Batista-Rodríguez JA, Pérez-Flores MA, Urrutia-Fucugauchi J. Three-dimensional gravity modeling of Chicxulub Crater structure, constrained with marine seismic data and land boreholes. Earth, Planets and Space. 2013;**65**:973-983

[8] Hinze WJ, Von Frese RRB, Saad AH. Gravity and Magnetic Exploration: Principles, Practices and Applications. New York: Cambridge University Press; 2013

[9] Long LT, Kaufmann RD. Acquisition and Analysis of Terrestrial Gravity Data. New York: Cambridge University Press; 2013

[10] Mehanee SA. Accurate and efficient regularized inversion approach for the interpretation of isolated gravity anomalies. Pure and Applied Geophysics. 2014;**171**:1897-1937

[11] Essa KS. New fast least-squares algorithm for estimating the best-fitting parameters due to simple geometric-structures from gravity anomsalies. Journal of Advanced Research. 2014;**5**(1):57-65

[12] Rama Rao BSR, Murthy IVR. Gravity and Magnetic Methods of Prospecting. New Delhi, India: Arnold-Heinemann Publishers; 1978. p. 390

[13] Chakravarthi V, Sundararajan N. Ridge regression algorithm for gravity inversion of fault structures with variable density. Geophysics. 2004;**69**:1394-1404

[14] Nettleton LL. Gravity and magnetics for geologists and seismologists. AAPG Bulletin. 1962;**46**:1815-1838

[15] Odegard ME, Berg JW. Gravity interpretation using the Fourier integral. Geophysics. 1965;**30**:424-438

[16] Sharma B, Geldart LP. Analysis of gravity anomalies of two-dimensional faults using Fourier transforms. Geophysical Prospecting. 1968;**16**:77-93

[17] Hartman RR, Teskey DJ, Friedberg JL. A system for rapid digital aeromagnetic interpretation. Geophysics. 1971;**36**:891-918

[18] Jain S. An automatic method of direct interpretation of magnetic profiles. Geophysics. 1976;**41**:531-541

[19] Thompson DT. EULDPH—A new technique for making computer-assisted depth estimates from magnetic data. Geophysics. 1982;**47**:31-37

[20] Gupta OP. A least-squares approach to depth determination from gravity data. Geophysics. 1983;**48**:357-360

[21] Kilty KT. Werner deconvolution of profile potential field data. Geophysics. 1983;**48**: 234-237

[22] Lines LR, Treitel S. A review of least-squares inversion and its application to geophysical problems. Geophysical Prospecting. 1984;**32**:159-186

[23] Bowin C, Scheer E, Smith W. Depth estimates from ratios of gravity, geoid, and gravity gradient anomalies. Geophysics. 1986;**51**:123-136

[24] Mohan NL, Anandababu L, Roa S. Gravity interpretation using the Mellin transform. Geophysics. 1986;**51**:114-122

[25] Nandi BK, Shaw RK, Agarwal NP. A short note on identification of the shape of simple causative sources from gravity data. Geophysical Prospecting. 1997;**45**:513-520

[26] Elawadi E, Salem A, Ushijima K. Detection of cavities and tunnels from gravity data using a neural network. Exploration Geophysics. 2004;**32**:204-208

[27] Salem A, Ravat D. A combined analytic signal and Euler method (AN-EUL) for automatic interpretation of magnetic data. Geophysics. 2003;**68**:1952-1961

[28] Essa KS. A simple formula for shape and depth determination from residual gravity anomalies. Acta Geophysica. 2007;**55**:182-190

[29] Asfahani J, Tlas M. An automatic method of direct interpretation of residual gravity anomaly profiles due to spheres and cylinders. Pure and Applied Geophysics. 2008;**165**:981-994

[30] Abedi M, Afshar A, Ardestani VE, Norouzi GH, Lucas C. Application of various methods for 2D inverse modeling of residual gravity anomalies. Acta Geophysica. 2009; **58**:317-336

[31] Asfahani J, Tlas M. Fair function minimization for direct interpretation of residual gravity anomaly profiles due to spheres and cylinders. Pure and Applied Geophysics. 2011;**168**:861-870

[32] Essa KS. A fast interpretation method for inverse modelling of residual gravity anomalies caused by simple geometry. Journal of Geological Research. 2012;**2012** Article ID: 327037

[33] Tarantola A. Inverse Problem Theory and Methods for Model Parameter Estimation. Philadelphia: SIAM; 2005

[34] Essa KS. Gravity data interpretation using the s-curves method. Journal of Geophysics and Engineering. 2007;**4**(2):204-213

[35] Kennedy J, Eberhart R. Particle swarm optimization. In: IEEE International Conference on Neural Networks (Perth, Australia). Vol. IV. Piscataway, NJ: IEEE Service Center; 1998. pp. 1942-1948

[36] Cedeno W, Agrafiotis DK. Using particle swarms for the development of QSAR models based on K-nearest neighbor and kernel regression. Journal of Computer-Aided Molecular Design. 2003;**17**:255-263

[37] Wachowiak MP, Smolíková R, Zheng Y, Zurada JM, Elmaghraby AS. An approach to multimodal biomedical image registration utilizing particle swarm optimization. IEEE Transactions on Evolutionary Computation. 2004;**8**:289-301

[38] Boeringer DW, Werner DH. Particle swarm optimization versus genetic algorithms for phased array synthesis. IEEE Transactions on Antennas and Propagation. 2004;**52**:771-779

[39] Chau WK. Application of a particle swarm optimization algorithm to hydrological problems. In: Robinson LN, editor. Water Resources Research Progress. New York: Nova Science Publishers Inc.; 2008. p. 3-12

[40] Eberhart RC, Shi Y. Particle swarm optimization: Developments, applications and resources. In: Proceedings of the Congress on Evolutionary Computation, Seoul, Korea; 2001. pp. 81-86

[41] Sweilam NH, El-Metwally K, Abdelazeem M. Self potential signal inversion to simple polarized bodies using the particle swarm optimization method: A visibility study. Journal of Applied Geophysics. 2007;**6**:195-208

[42] Parsopoulos KE, Vrahatis MN. Recent approaches to global optimization problems through Particle Swarm Optimization. Natural Computing. 2002;**1**:235-306

[43] Werner S. Interpretation of magnetic anomalies at sheet-like bodies. Sveriges Geologiska Undersok, Series C, Arsbok. 1953;**43**(6)

[44] Abdelrahman EM, Bayoumi AI, Abdelhady YE, Gobashy MM, El-Araby HM. Gravity interpretation using correlation factors between successive least squares residual anomalies. Geophysics. 1989;**54**:1614-1621

[45] Said R. The Geology of Egypt. Amsterdam: Elsevier; 1962

[46] Beadnell HJL. The Cretaceous Region of Abu Roash, near the Pyramids of Giza, Egypt. Cairo: Survey Department; 1902. p. 48

[47] Faris MI. Contributions to the stratigraphy of Abu Roash and the history of the upper Cretaceous in Egypt. Bulletin of Faculty of Science, Cairo University. 1948;**27**:221-239

[48] Abdelrahman EM, El-Araby HM. Shape and depth solutions from gravity data using correlation factors between successive least-squares residuals. Geophysics. 1993;**59**:1785-1791

Chapter 2

3D Modeling and Inversion of Gravity Data in Exploration Scale

Hongzhu Cai, Bin Xiong and Yue Zhu

Additional information is available at the end of the chapter

http://dx.doi.org/10.5772/intechopen.70961

Abstract

The gravity method has been widely used for detecting the subsurface density anomaly and geological structures. The interpretation result based on gravity data can be used for mineral/oil exploration and regional geological study. The effective and successful application of gravity methods depends on the fast forward modeling and stable inversion tools to image the subsurface density structures. In this chapter, we will review the applications and developments of gravity method. We starts form the basic theory for gravity field and the scalar gravity potential and introduce the closed form of the solution for the gravity field caused by a density anomaly. Different gravity data forward modeling and inversion techniques will be introduced in this chapter with their application in petroleum reconnaissance. Several examples will be presented in this chapter to illustrate the application of different gravity modeling and inversion techniques.

Keywords: modeling, inversion, density, depth, imaging

1. Introduction

Gravity method is an important geophysical tool for subsurface mapping of the density distributions. The development of high-accuracy instrumentation and stable observation platform since 1980s has made the large scale airborne gravity survey become possible. The interpretation result based on gravity data can be used for mineral and oil exploration. For example, the gravity data has been used to estimate the depth of the sedimentary basin and detect sub-sea basalt which is usually associated with the oil reservoir.

The gravity modeling has its origin back to Newton's law of universal gravitation developed in 1687, which is equivalent to solving a Poisson's equation for scalar gravity potential. One can use different approaches to solve this Poisson's equation, e.g., finite difference or finite element method. However, the most common approach is to solve it in an integral way using Green's function. In the first section of this chapter, we will review the Poisson's equation and its integral solution for a general three-dimensional case. The numerical approaches to evaluate the integrals in the solution of the Poisson's equation will be discussed. For more realistic problems, we also provide an approach of gravity modeling to take into account of complex geometry such as topography.

Conventional gravity inversion is based on discretizing the earth model into a set of rectangular prisms, and considering each prism has a constant density. Given observed gravity and/or gravity gradiometry fields, one is interested to recover the density within each cell. Similar as other geophysical inverse problems, this volumetric inversion of gravity data is an ill-posed problem which is strongly affected by the non-uniqueness and instability. To solve this ill-posed a problem, we will first introduce an objective functional which contains two parts, the misfit functional for least-square fitting of the observed data, and the stabilizer for regularizing the subsurface density distributions. The solutions for solving the minimization problem of this objective functional will then be briefly discussed. Generally, the volumetric gravity inversion produces diffusive images of the subsurface density distribution with very limited depth resolution (strictly speaking, the gravity method, with observation above the earth's surface, does not have any depth resolution for the subsurface density distribution). One may improve the gravity inversion results by adding more constraints, e.g., logarithmic barrier to constrain upper and lower limits of the density, focusing stabilizer to produce sharper boundaries between the density contrasts, and etc., in the inversion.

A more advanced approach of gravity inversion is to invert for the shape of the anomaly instead of the three-dimensional volumetric density distribution. In certain applications, such as sedimentary basin analysis, it is more interesting to estimate the location of the sediment-basement interface and the shape by assuming that the density distribution of the sediments in vertical direction is known based on some other a priori information such as well logging. To solve this specific problem, the geometric inversion method to recover the depth to basement has been proposed in recent years. These types of methods are mostly based on the discretization of the sedimentary columns or the sediment-basement interface instead of the 3D subsurface. By adopting this new discretization, the parameters of unknown in the inversion and the model uncertainty can be reduced significantly. In the last section of this chapter, we will discuss the most updated geometric modeling and inversion method.

2. Three-dimensional gravity modeling

2.1. Basic theory for gravity method

In this section, we will introduce the basic gravity theory and the corresponding solution for solving the gravity modeling problem. We start from the basic Poisson's equation for gravity problem and introduce the solution for gravity field data caused by some mass. For numerical solution, we show the discretized formulae for calculating the gravity field data. We also briefly discuss the fast modeling methods which is based on fast Fourier transform (FFT). In certain application, the gravity problem can be reduced to study the response caused by a

http://dx.doi.org/10.5772/intechopen.70961

sediment-basement interface model (or it can be called as the density contrast model). In this case, the forward modeling problem can be solved using the column discretization or the surface discretization method based on 3D analog of Cauchy-type integral.

We first consider a simple gravity problem as shown in **Figure 1** which contains a domain filled by some density distribution of $\rho(r)$. It can be shown that the gravity field caused by this density distribution satisfies the following equation [1–3]:

$$\nabla \cdot g = -4\pi\gamma\rho, \quad \nabla \times g = 0, \tag{1}$$

where the parameter γ denotes the gravitational constant. We can see from Eq. (1) that the gravity field g is curl free which is also referred as a conservative vector field. Take into consideration of the fact that $\nabla \times (\nabla U) = 0$, one can construct the solution for Eq. (1) by introducing the scalar gravity potential U such that:

$$g = \nabla U. \tag{2}$$

As a result, the gravity field satisfies the Poisson's equation [1, 2]:

$$\nabla^2 U = -4\pi\gamma\rho. \tag{3}$$

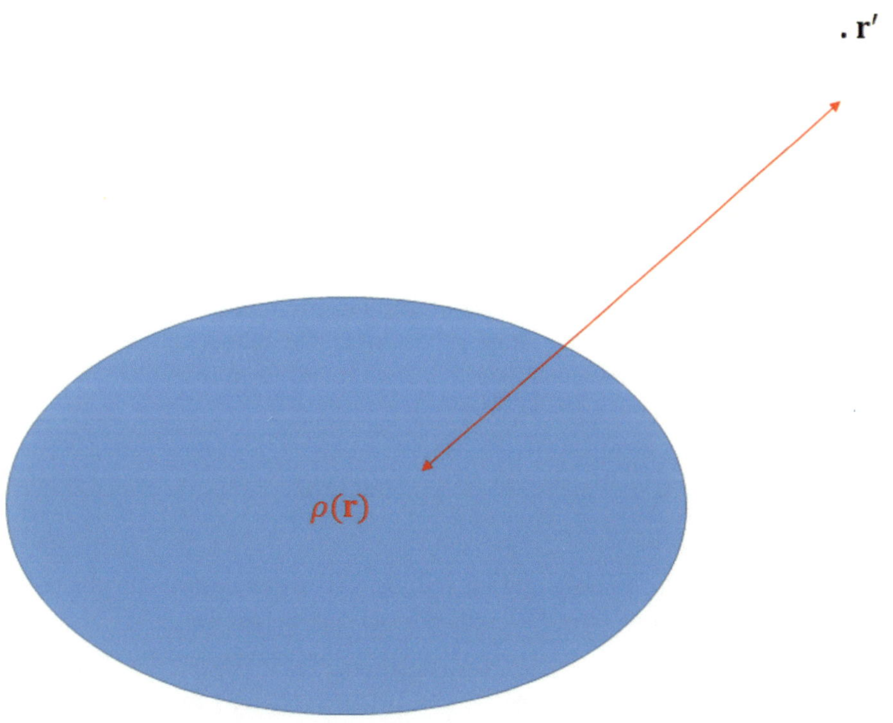

Figure 1. Illustration for gravity problem.

2.2. Integral representation of gravity anomaly

For geophysical problem, the observation point is usually outside the mass, in this case, Eq. (3) can be reduced to the well-known Laplace equation. In this case, the solution for gravity field outside the source domain can be written as:

$$g(r') = \gamma \iiint_D \rho(r) \frac{r - r'}{|r - r'|^3} dv, \tag{4}$$

where D indicates the domain with anomalous mass. The corresponding Green's function for the gravity field caused by a point source can be written as follows:

$$G(r, r') = \frac{r - r'}{|r - r'|^3}. \tag{5}$$

By comparing Eqs. (4) and (5), one can see that the gravity field caused by some mass with arbitrary shape can be calculated by integral the solution for the point source.

By taking the spatial derivative of Eq. (4), one can find the integral representation of the gravity gradiometry data which is defined as:

$$g_{\alpha\beta} = \begin{bmatrix} g_{xx} & g_{xy} & g_{xz} \\ g_{yx} & g_{yy} & g_{yz} \\ g_{zx} & g_{zy} & g_{zz} \end{bmatrix}. \tag{6}$$

It can be proved that such gravity tensor is symmetric and the summation of the diagonal components equals to zero (this is because scalar gravity potential is Laplacian: $\nabla^2 U = g_{xx} + g_{yy} + g_{zz} = 0$, outside the source region). As a result, among these nine gravity tensor components, only five of them are independent.

The integral formula in Eq. (4) and its derived formulae are the common approach for calculating the gravity anomaly caused by some excess mass. For numerical modeling and inversion of gravity data, it is convenient to write a discretized form of Eq. (4) by dividing the subsurface mass anomaly into a grid of prism cells:

$$g(r'_n) = \gamma \sum_{k=1}^{N_m} g_k \rho_k \Delta x \Delta y \Delta z, \tag{7}$$

where ρ_k denotes the density value of the kth cell; Δx, Δy, Δz denotes the dimension of the cell in x, y, z direction; g_k is the kernel function which is defined as follows:

$$g_k = \frac{r_k - r'_n}{|r_k - r'_n|^3}. \tag{8}$$

Numerically, Eq. (8) can be evaluated using the Gaussian quadrature or the simplest central point integral formula which could be less accurate if the receiver is closer to the cell.

2.3. Fast Fourier transform (FFT) method for calculating gravity anomaly

In practical application, the modeling domain can be very large which may result in millions of cells. In this scenario, the calculation of gravity field directly using Eqs. (4) and (7) can be time and memory consuming.

Geophysicists have attempted, for decades, to apply the fast Fourier transform algorithm to potential field data modeling and inversion [4]. In the pioneering work of Parker and Oldenburg [5, 6], it has been shown that the FFT method can be used to calculate the gravity field caused by a density model contains different non-flat layers. Nagendra et al. [7] have released a FORTRAN code for gravity modeling which is based on the method of Parker and Oldenburg [5, 6].

The more advanced method for calculating gravity anomaly caused by complex geometry can be found in [8]. From the numerical perspective, the whole density anomaly domain is divided into a series of horizontal layers extends infinitely in x and y direction. For each layer, the corresponding gravity anomaly is calculated using FFT method and the results for each layer will be summed together to get the total gravity response. This approach usually requires a uniform discretization in x and y direction. Interested readers are recommended to get more detailed work in [8].

2.4. Differential equation for gravity modeling

Another alternative approach to overcome the large memory and computation requirement, which occurs in the analytical integral representation, is the differential equation method which includes finite difference, finite element and finite volume methods. These types of methods solve the Poisson's equation in Eq. (3) directly with the homogeneous Dirichlet boundary condition. All these method finally result in the sparse system of equation for the scalar gravity potential. Farquharson and Mosher [9] implemented a 3D finite difference algorithm for solving the Poisson's equation of scalar gravity potential. The numerical study has shown its high accuracy (relative error is around 1%) comparing to the analytical method with direct integral over each cell. This method runs slower than the conventional integral method but it uses much less memory.

The finite element or finite volume method with unstructured mesh (e.g., tetrahedral element for three-dimensional case), can be used for solving this problem with much less memory consumption [10]. In this chapter, we will take the finite element method with unstructured tetrahedral mesh for example. We consider a tetrahedral element shown in **Figure 2** with node indexing [11]. We assume that in each tetrahedral element, the scalar gravity potential is defined in the node. The gravity potential inside the tetrahedral element can be written as follows:

$$U_e(x, y, z) = \sum_{j=1}^{4} N_e^j(x, y, z) U_e^j, \tag{9}$$

where $N_e^j(x, y, z)$ is the linear basis function for the tetrahedral element [11].

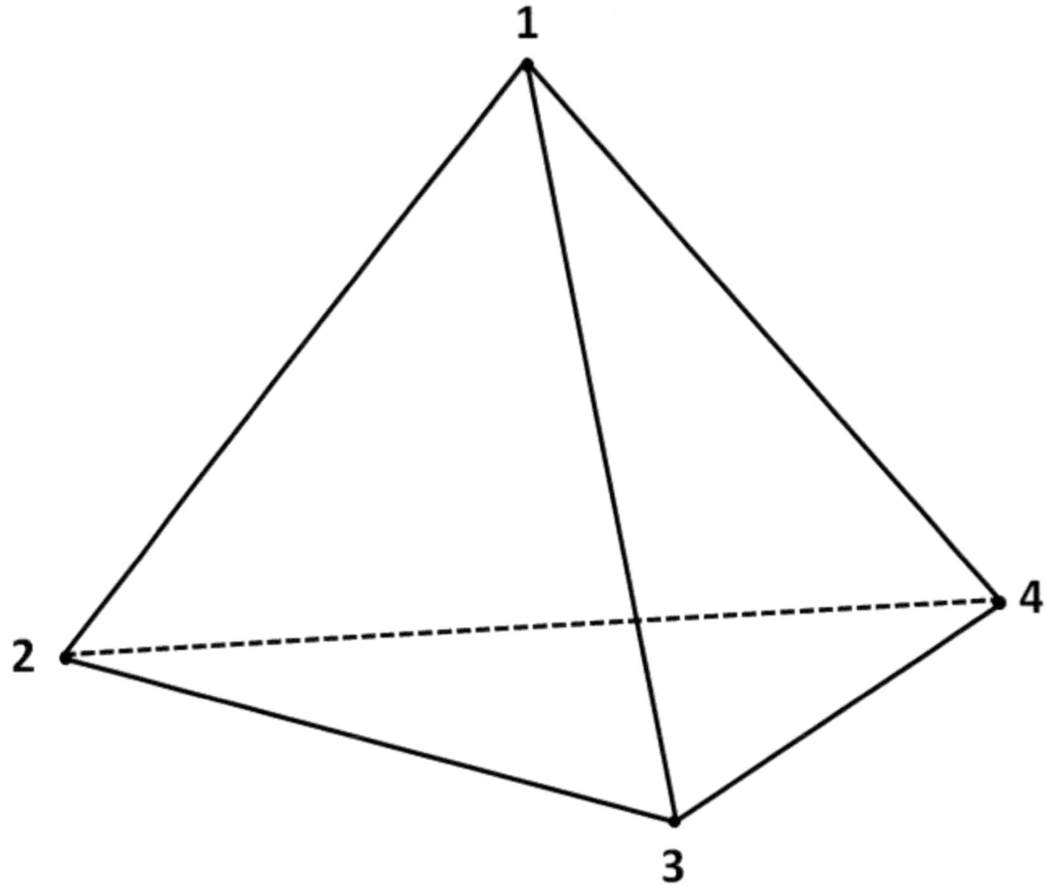

Figure 2. An illustration for tetrahedral element with node indexing.

After applying finite element analysis to the Poisson's equation for scalar gravity potential, one can find a sparse system of equations as follows:

$$KU = b, \tag{10}$$

where K is the finite element stiffness matrix [11], b is the source term which is related with the density distribution. After applying the proper boundary condition (e.g., the homogeneous Dirichlet boundary condition), the system of equation in Eq. (10) can be solved either with iterative method or the modern direct solvers. It has been demonstrated that that finite element method for gravity modeling can simulate complex geological structures. Furthermore, the method is advantageous to the conventional integral method in terms of both memory and computation speed for realistic large scale geological models.

Figure 3 shows an example of how powerful of finite element with unstructured tetrahedral mesh for simulating complex geological structures. The model represents the Voisey's Bay

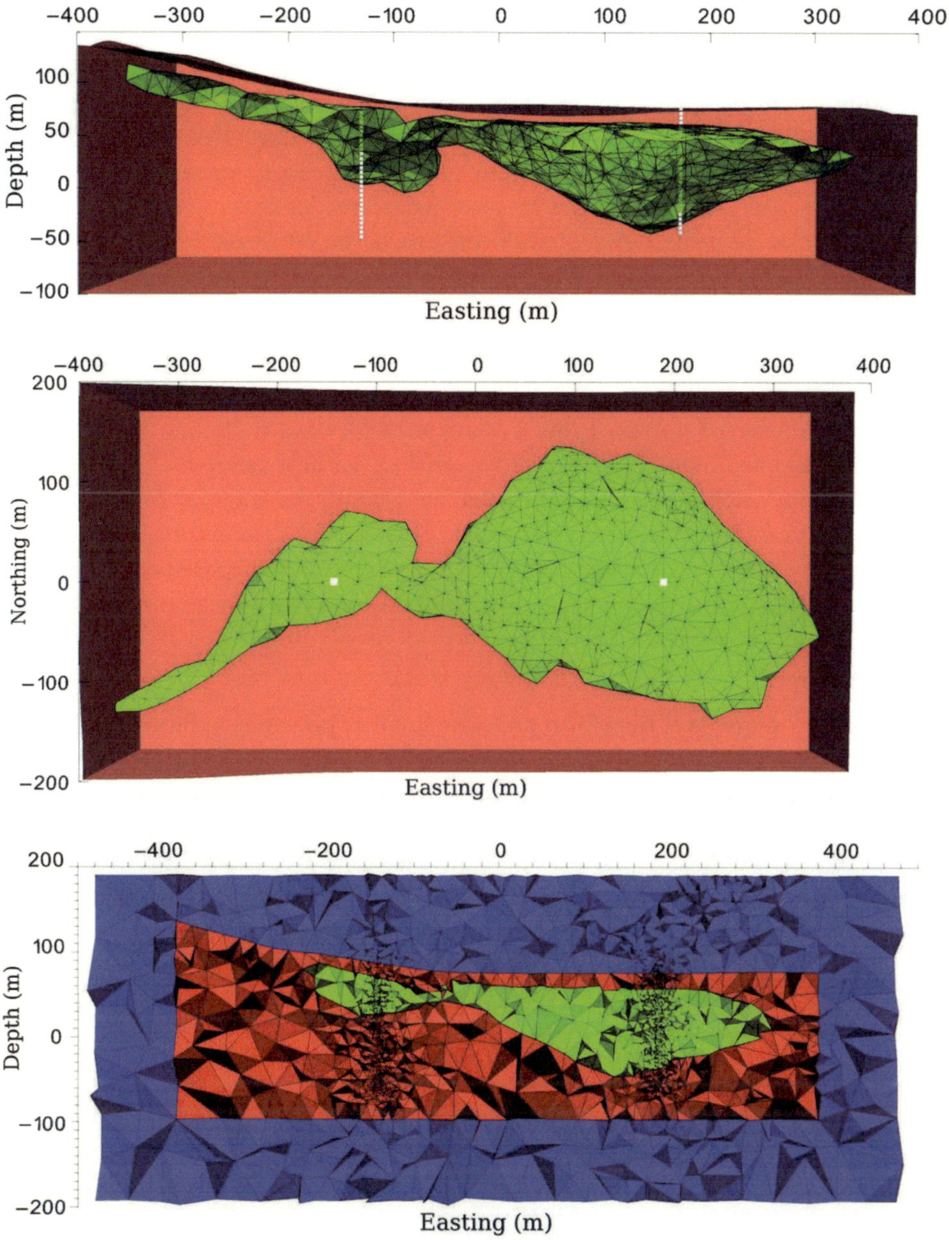

Figure 3. An example of simulating complex geological models (Voisey's Bay deposit) using finite element method with tetrahedral mesh. The upper and middle panels show the side and plane views, while the bottom panel shows a vertical section [10].

massive sulfide deposits located in Labrador, Canada [10]. One can see that the mesh is locally refined nearby the area with complex geometry. It is difficult to simulate such complex model using the regular prism grid.

2.5. Advanced method for the modeling of density interface model

In oil and gas exploration, it is more interesting recover the favoring geological structure for oil/gas accumulation, using gravity method. Usually, the density of the basement rocks is relatively higher than the sediment density due to the compaction effect during the sedimentation process. In this scenario, the variation of the depth to basement can cause gravity anomaly on the earth's surface and can be recorded. Consider that the density of sediments and basement rocks are well known or well constrained by other geological information, we can use the observed gravity anomaly to estimate the depth to the crystalline basement. Such research topic has been investigated for decades [12–18].

Figure 4 shows an illustration of a synthetic sediment-basement interface model where the sedimentary rock and the basement rock are characterized by different density. The classic and most straightforward approach for solving such forward modeling problem is based on discretizing the sedimentary pack into a grid of vertical columns. The gravity field anomaly caused by each column can be calculated using the integral formula in Eq. (4). Such expressions can be further reduced in the special case of constant sediment density which results in a constant density contrast along the sediment-basement interface. In a special case of density contrast change exponentially or quadratically with depth, there exists other special formulation for the forward modeling. Interested readers can refer to [14] for more details.

The elegant theory of integral transform provides another powerful approach for solving the gravity forward modeling problem using 3D analogy of Cauchy-type integral [1]. We consider a density contrast model shown in **Figure 5**. The reference model with two layers are separated by a horizontal plane P (at $z = -H_0$), which is a density contrast interface. For the idealized two-layered model, the density above and below the plane P are two different constants. In real case, the actual density contrast interface Γ is an arbitrary surface. As a result, the domain D_R, which is bounded by plane P and surface Γ can cause some gravity anomaly.

Using the integral transform approach [1], such gravity anomaly can be represented as follows:

$$g(r') = -4\pi\gamma\Delta\rho C^{\Gamma_R}(r', h(x,y)d_z), \tag{11}$$

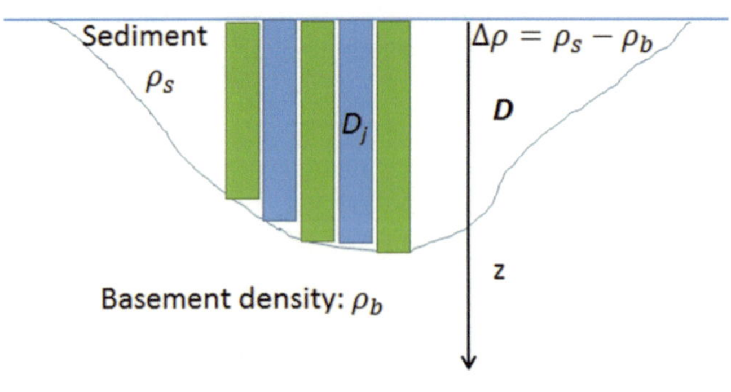

Figure 4. A typical sediment-basement interface model.

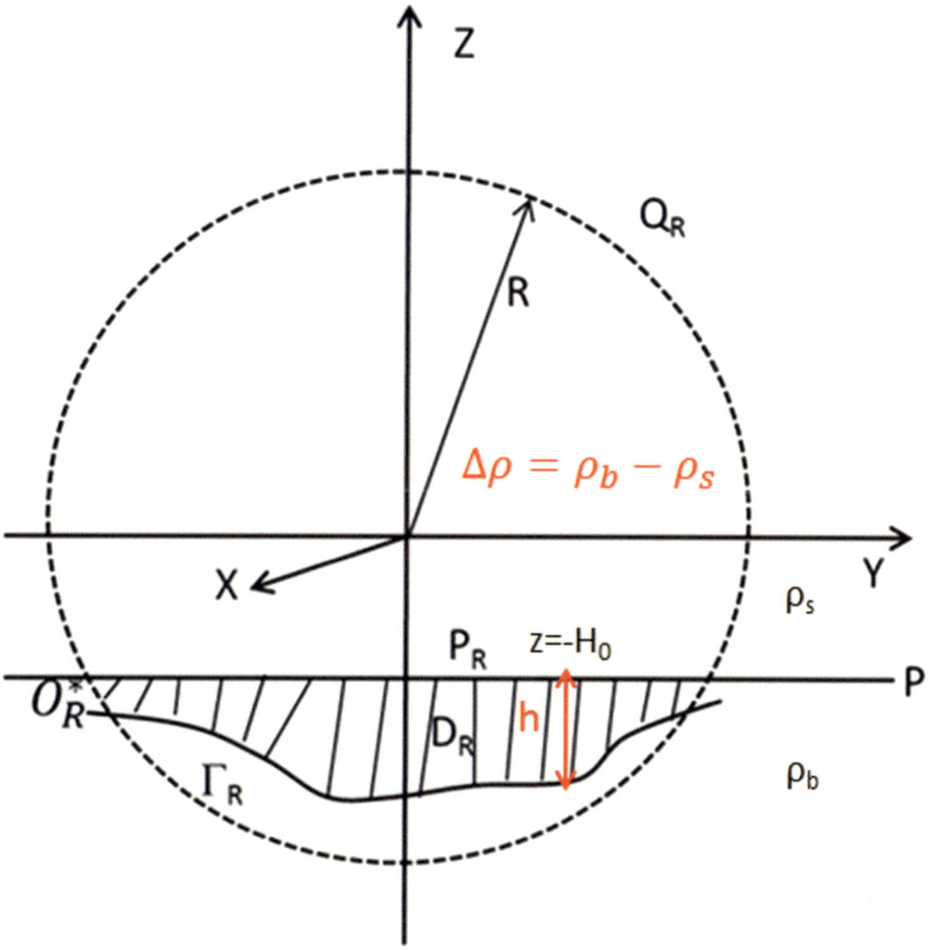

Figure 5. Illustration of density contrast model for Cauchy-type integral representation.

where $h(x, y)$ is the relative elevation between the surface Γ and plane P at each horizontal location, $\Delta\rho$ is the density contrast between these two layer which is also the anomalous density inside domain D_R, $C^{\Gamma_R}(r', h(x, y)d_z)$ is the 3D Cauchy-type integral on the surface of Γ_R for the vector function $h(x, y)d_z$, at the point of r'.

Mathematically, the 3D analog of Cauchy-type integral for the vector function $\phi(r)$ can be defined as follows [1, 19, 20]:

$$C^S(r', \phi) = \frac{-1}{4\pi} \iint_S \left[(n \cdot \phi) \nabla \frac{1}{|r - r'|} + (n \times \phi) \times \nabla \frac{1}{|r - r'|} \right], \tag{12}$$

where S is some closed surface which bounds a 3D domain D, $\phi(r)$ is some vector function defined on the surface S. The 3D Cauchy-type integral itself is a vector function which satisfies the following equation outside the domain D which is bounded by surface S [1, 19, 20]:

$$\nabla \cdot C^s = 0, \nabla \times C^s = 0. \tag{13}$$

These important properties make it possible for using the 3D analog of Cauchy-type integral to construct solutions of gravity problem. Using the scalar representation of 3D Cauchy-type integral, one can write the solution of gravity field caused by the model shown in **Figure 5** as follows [1, 19, 20]:

$$g_\alpha = -\gamma\Delta\rho \iint_{P_R} \Delta_{azy\eta} \frac{h(x,y)\left(r_\eta - r'_\eta\right)}{|r - r'|^3} b_\gamma dxdy, \tag{14}$$

where each of α, γ, η can be equal to x, y, z; P_R is the projection of Γ_R on the horizontal plane P. The four-index Δ symbol is defined as follows [1, 19, 20]:

$$\Delta_{azy\eta} = \delta_{\alpha z}\delta_{\gamma\eta} + \delta_{\alpha\eta}\delta_{z\gamma} - \delta_{\alpha\gamma}\delta_{z\eta}; \delta_{\alpha\beta} = \begin{cases} 1, \alpha = \beta \\ 0, \alpha \neq \beta \end{cases}, \tag{15}$$

the parameter b_γ is defined as follows:

$$b_x = \frac{\partial h}{\partial x}, b_y = \frac{\partial h}{\partial y}, b_z = -1. \tag{16}$$

Note that the model described in **Figure 5** is exactly a typical sediment-basement interface model when the horizontal plane P is on the earth's surface. By taking the spatial derivative of Eq. (14), one can find the expression of gravity gradiometry data using 3D analog of Cauchy-type integral transform. Clearly, we can see that the 3D gravity modeling problem can be reduced to the surface integral over the density contrast surface using the Cauchy-type integral approach. As a result, the computational cost can be reduced significantly comparing to the direct integral method. As a matter of fact, this method works for a complex distribution of density contrast over vertical direction. In this case, it only requires the density contrast function in vertical direction to be integral [1, 19, 20].

3. Three-dimensional gravity inversion

As we know, the inversion of gravity data is a serious non-unique problem. In a general case, the model parameter is much larger than the observed data points. There exists infinity number of models which can fit the gravity data in a least-square sense [2]. Furthermore, the potential filed data does not have any depth resolution. In order to obtain the most reasonable solution with correct depth resolution, it is crucial to apply regularization and some a priori information during the inversion process. In this section, we will briefly discuss some modern 3D inversion techniques for gravity field data. First of all, we will formulate the inverse problem for the 3D volumetric inversion and introduce some method for minimizing the objective/parametric functional. Follow this, we will discuss the geometric inversion approach for the depth to basement inversion or density contrast interface inversion. Finally, we will introduce the joint inversion approach.

3.1. Conventional volumetric inversion

One of the most important applications of gravity inversion is to recover the 3D subsurface density distribution. The conventional approach for solving this problem is based on discretizing the subsurface into a grid of 3D prism cells. During the inversion, the density values for each cell will be adjusted in order to fit the observed gravity data. In order to fit the data with a reasonable density model, one can construct a parametric functional defined as follows [2]:

$$P(d,m) = (W_d A(m) - W_d d)^T (W_d A(m) - W_d d) + \alpha \left(W_m m - W_m m_{apr}\right)^T \left(W_m m - W_m m_{apr}\right) \quad (17)$$

where d is the vector for observed data, m is the vector for model parameters, m_{apr} represents the a priori model, A is the forward modeling operator which is linear in this case, W_d is the data weighting matrix, W_m is the model weighting matrix, and α is the regularization parameter.

In practical application, a smaller regularization parameter can cause over fitting for the data and result in some unreasonable model. A larger value of regularization parameter will place strong penalty and enforce the model parameter be closer to the a priori model with the price of a bad data fitting. Several methods have been proposed for choosing the optimized regularization parameter. These methods include, but not limited to, the adaptive selection method and the L-curve method [2, 21].

The model weighting matrix can be calculated based on integral sensitivity method for reasonable depth resolution [2]. In this case, the second term in the right hand side of Eq. (17) is equivalent to the minimum norm regularization which usually produce a smooth density distribution for 3D gravity inversion.

The minimization of the parametric functional in Eq. (17) can be solved using gradient type method such as steepest ascend method, conjugate gradient method and Newton method. In the gradient type inversion, one has to obtain the sensitivity matrix. For volumetric gravity inversion problem, the sensitivity matrix is exactly the same as the forward modeling matrix. For large scale inversion, the moving sensitivity domain approach is usually used to reduce the computation cost and memory consumption [22]. Within the framework of this approach, only the model cells within some distance from the receivers will be considered. If the cells are far away, the sensitivity is assumed to be zero. As a result, the full sensitivity matrix is reduced to a sparse matrix.

Usually, the inversion of gravity data in the original linear model space can result in unreasonable value of density distribution due to the absence of constraints. In reality, we usually know the estimated density values or their upper and lower boundaries. In this case, it is desirable to transfer the model parameter m in the original linear model space into the model parameter $\tilde{m}$ in logarithmic space as follows [2, 23]:

$$\tilde{m} = \ln\left(\frac{m - m_1}{m_2 - m}\right), \tag{18}$$

where m_1 and m_2 are the lower and upper limits of model parameter m. The inversion is done in the logarithmic model space of $\tilde{m}$ and will be transformed back to the original linear model space. It can be shown that such transformation can guarantee the inverted model parameter value for m to be in the range defined by $[m_1, m_2]$.

However, the gradient type method for the deterministic inversion is characterized by the local minimum problem, which means that the solution is dependent on the selection of initial model [24]. If the initial model is not chosen properly or closer to the true model, the inverted model could be some local minimum instead of the global minimum that we usually expected. In order to solve this problem, the methods such as Monte Carlo method, genetic algorithm, simulated annealing method can be used to reach the global minimum [24]. Comparing to the gradient type method, these methods search the optimized model from the whole model space for the global minimum. However, these methods are usually time consuming, especially for full 3D inversion which may contain millions of model parameters.

Instead of the deterministic inversion, the stochastic inversion approach can be applied to the potential field data [24–26]. Comparing to the deterministic inversion, the stochastic inversion can also provide uncertainty estimation for the model confidence. Recently, we start to see more publications in geophysical stochastic inversion for potential field data.

3.2. Density contrast interface inversion

The conventional prism based inversion usually produce diffusive images even though some techniques such as focusing inversion [27] can be used to enforce a sharp boundary between different geological units. In petroleum reconnaissance using gravity method, it is of great importance to estimate the depth to basement. However, from the interpretation of conventional prism based inversion, it is difficult to pick up the correct location of such interface due to the non-uniqueness of the inversion and the low resolution of inverted density distribution. In such environment, the density contrast between sedimentary rocks and basement is usually well known based on other information such as drilling. The gravity anomaly can be attributed to the variation of the sediment-basement interface. Based on the method introduced in Section 2, this type of models can be simulated using the column discretization or the Cauchy-type integral approach. In this subsection, we will mainly focus on the inversion of sediment-basement interface using the 3D Cauchy-type integral approach.

Within the framework of this approach, we formulate the inversion with respect to the depth to basement and the density contrast value (may not necessary be a constant). Similar to the prism based inversion, we can formulate the inverse problem using the parametric functional introduced in Eq. (17). However, the model parameters now become as follows [19]:

$$m = [h; \rho], \tag{19}$$

where h is the depth to basement at different horizontal locations, and ρ is the density contrast value between the sediment and basement. Comparing to the prism based inversion for the density distribution, the forward modeling operator A which is related with the 3D analog of Cauchy-type integral, for density contrast surface inversion is a nonlinear operator since the gravity data does not have a simple relationship with the depth. Such inversion can be solved with the gradient type inversion methods. The sensitivity matrix for the depth to basement and the density contrast values can be calculated by directly differentiate Eq. (14) with respect to the depth to basement and the density contrast value. In some applications, the density contrast values are usually well known based on well logging. Under this circumstance, only the depth to basement value need to be inverted and the non-uniqueness of the inversion can be reduced significantly.

As an example, we consider a 3D sedimentary interface model with the vertical section shown in **Figure 6**. We consider a constant density value for the basement rock. To be realistic, we assume that the density value for sediment increase exponentially with depth due to compaction. As a result, the density contrast value will decrease exponentially with depth and approach the basement density which is assumed to be a constant. The synthetic gravity data is simulated on the earth's surface and will be used for inversion. For this model, we consider that the density contrast profile with depth is already known and only invert for the depth to basement. **Figure 7** shows a comparison between the true model and the inverted model using

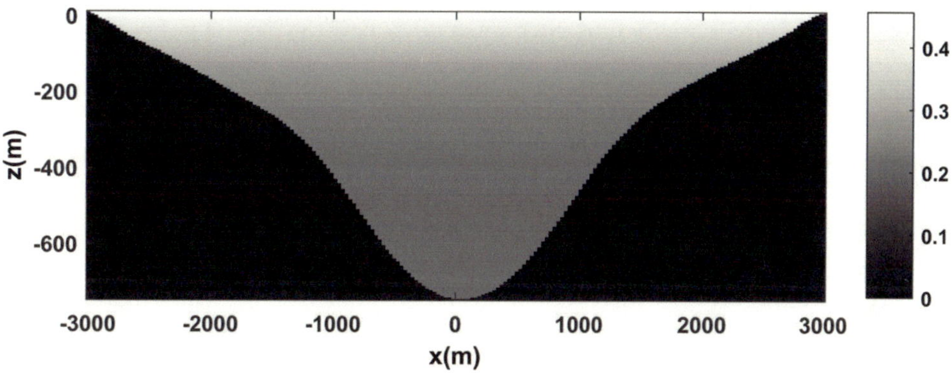

Figure 6. A vertical section (y = 0) of the sediment-interface model with exponential density contrast in vertical direction [19].

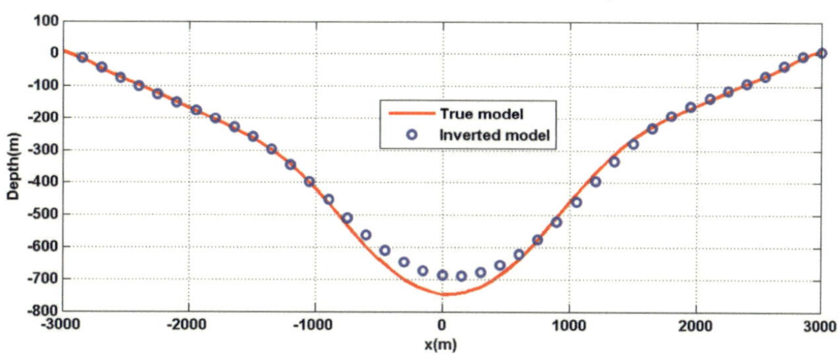

Figure 7. Inversion result for the synthetic sediment-basement interface model with exponential density contrast [19].

the Cauchy-type integral approach. One can see that the depth to basement and the shape of the sedimentary basin is well recovered in the inversion.

During the inversion, we use a flat surface as the initial model and a priori model. Actually, such inversion method is robust enough and does not depend too much on the selection of initial model and a priori model. However, one can use some other initial model in order to speed up the convergence of the inversion. The famous Bouguer slab formula [28] can be used as an initial model:

$$h = \frac{g_B \Delta\rho_0}{41.89\Delta\rho_0 + a g_B} \tag{20}$$

where g_B denotes the observed Bouguer gravity anomaly, $\Delta\rho_0$ is the density contrast value on the earth's surface and a is the gradient of density contrast in vertical direction. As one may note, this equation works properly for the constant density contrast or linear density contrast but it does provide a good approximation of the depth to basement for a general case, e.g., exponentially increased density contrast profile with depth.

3.3. New developments in gravity inversion

In this subsection, we will briefly introduce some other new developed techniques for gravity inversion. These new methods include, but not limited to, the binary inversion, multinary inversion and the joint inversion approach.

3.3.1. Binary and multinary inversion of gravity data

In some application of gravity imaging such as subsurface tunnel detection, salt structure imaging, the density range of the target and the density of host rock are usually well known or well constrained [29]. However, the conventional inversion in this case will still produce a diffusive image with spread density ranges and continuous model space. In reality, the inverted density value should be clustered nearby the density value of the host rock and the density of the target.

In the case of a model with two distinct density values, the continuous inversion parameters m in the original model space can be transformed into a new binary space for inversion [30]. Zhdanov and Cox [29] introduced a multinary inversion approach for geological models with more than two density values for different geological units. Several different functions, such as Heaviside function and Gaussian function, can be used for multinary transformation [31]. Within the framework of multinary approach, the value of the inverted model parameters is enforced to be nearby the vicinities of the preselected values.

3.3.2. Joint inversion approach

As we know that each geophysical method and data is only sensitive to specific model parameter. Due to the inherent non-uniqueness of geophysical data inversion, the recovered model parameter from individual inversion can be ambiguous. Such ambiguity can be reduced by incorporating more a priori information into the inversion.

Different geophysical models can be coupled with each other either directly or structurally. For example, the density and seismic velocity can be related with each other by empirical equations. Alternatively, a density model can be related with the velocity model by assuming the structurally similarity. As a result, it is possible to enforce the coupling between different model parameters by inverting this different geophysical data set simultaneously. The structurally similarity based joint inversion can be achieved by minimizing the cross gradients between different model parameters [32–34]. Within the framework of this approach, the structural similarity, between model parameter m_1 and m_2, can be measured by the cross gradient which is defined as follows [32]:

$$\theta(x, y, z) = \nabla m_1 \times \nabla m_2. \tag{21}$$

Such regularization term will be minimized during the minimization of data misfit functional for the joint inversion approach.

Zhdanov [31] proposed a new and more flexible joint inversion approach based on Gramian constraint. Within the framework of this approach, different model parameters are coupled through the Gramian matrix which can either force the direct relationship between different model parameters or their spatial gradients. One good property of such joint inversion is that the algorithm will only enforce such coupling when it does exist and will not introduce artificial coupling when there is no relationship or coupling between different model parameters [31].

It is straightforward that the joint inversion formulation can be simplified if there exist a shared model parameter for different geophysical data set. For example, the DC electric method and the magnetotelluric (MT) method both invert for the electric conductivity. As a result, it is unnecessary to formulate the joint inversion using the approach that we have just discussed, in this scenario. Based on this idea, the joint inversion for depth to basement using different geophysical data can be greatly simplified by considering that the depth to basement is a shared model parameter for different data set such as gravity and MT data. In the meantime, each method may also have a private model parameter such as density contrast for gravity data and conductivity contrast for MT data.

Here we consider a synthetic sediment-basement interface model [19, 35]. The synthetic gravity and MT data are simulated on the earth's surface for this model. These data will be used to recover the sediment-basement interface. Furthermore, we assume that the density contrast value and conductivity contrast value are also unknown. Under this circumstance, the inversion of individual data set is characterized by strong non-uniqueness. By assuming the shared model parameter of depth to basement, for gravity and MT data, during the joint inversion approach, the recovered model parameters is much closer to their true value comparing to the individual data inversion (as one can see in **Figure 8**).

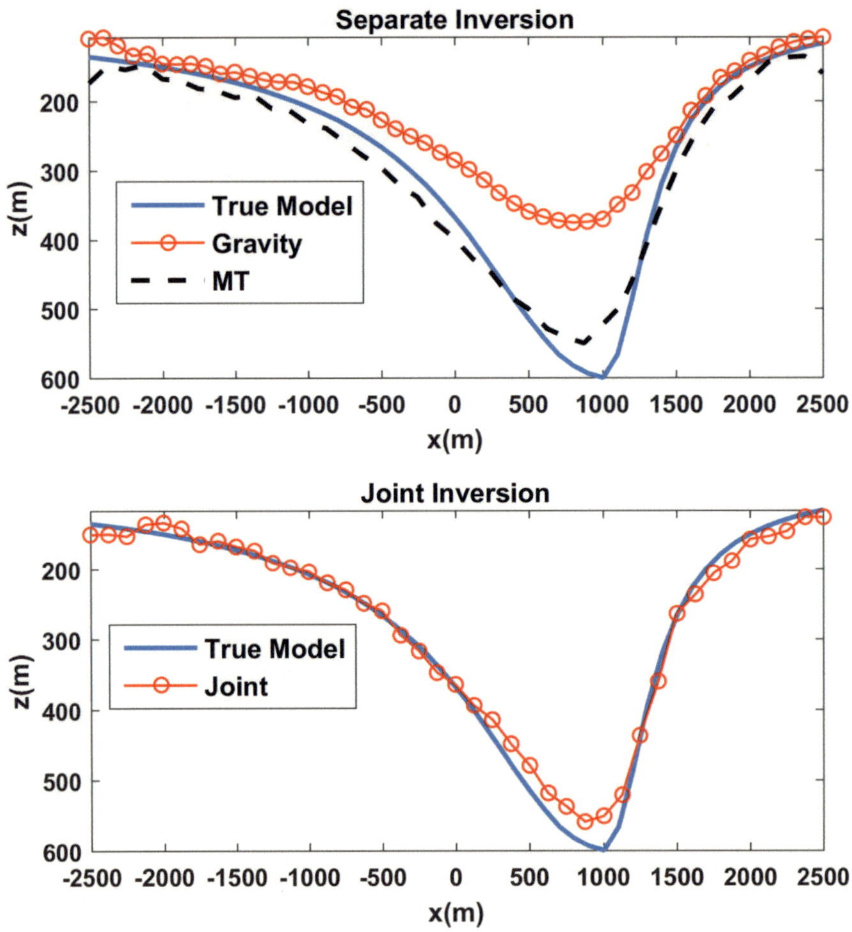

Figure 8. A comparison between separate gravity and MT inversion for depth to basement, and the joint inversion result for a synthetic sediment-basement interface model [19].

4. Summary

In this chapter, we have reviewed the 3D gravity forward modeling and inversion problem. We start from the Poisson's equation for scalar gravity potential and introduce the formula of gravity field in integral form based on the Green's function which corresponds to the solution of a point source. We also introduced some other techniques, such as FFT method and differential equation method, for fast and efficient solving of the gravity forward modeling problem. In the application of sedimentary basin modeling, we have introduced the column discretization method and the advanced Cauchy-type integral method. In the second section of this chapter, we introduced the gravity inversion problem and first start from the conventional prism based inversion. Following this, we introduce another application of gravity inversion for locating the density contrast interface. We mainly focus on the 3D inversion based on Cauchy-type integral for recovering the depth information. Finally, we have introduced some other new recently developed techniques for gravity inversion which includes, but not limits to, the multinary inversion and joint inversion approach. The interested readers are recommended to read the relevant publications cited in this chapter, for more details.

Acknowledgements

The authors acknowledge Dr. Jahandari and Professor Farquharson from the Memorial University of Newfoundland for releasing the useful figures in their publications.

Author details

Hongzhu Cai[1]*, Bin Xiong[2] and Yue Zhu[3]

*Address all correspondence to: caihongzhu@hotmail.com

1 Aarhus University, Aarhus, Denmark

2 Guilin University of Technology, Guilin, China

3 University of Utah, Salt Lake City, USA

References

[1] Zhdanov MS. Integral Transforms in Geophysics. Springer-Verlag; Berlin. 1988

[2] Zhdanov MS. Geophysical Inverse Theory and Regularization. Amsterdam: Elsevier; 2002

[3] Kaufman AA, Hansen RO. Principles of the Gravitational Method. Amsterdam: Elsevier; 2007

[4] Shin YH, Choi KS, Xu H. Three-dimensional forward and inverse models for gravity fields based on the Fast Fourier Transform. Computers & Geosciences. 2006;**32**(6):727-738

[5] Parker RL. The rapid calculation of potential anomalies. Geophysical Journal International. 1973;**31**(4):447-455

[6] Oldenburg DW. The inversion and interpretation of gravity anomalies. Geophysics. 1974;**39**(4):526-536

[7] Nagendra R, Prasad PVS, Bhimasankaram VLS. Forward and inverse computer modeling of a gravity field resulting from a density interface using Parker-Oldenberg method. Computers & Geosciences. 1996;**22**(3):227-237

[8] LaFehr TR, Nabighian MN. Fundamentals of Gravity Exploration. Tulsa: Society of Exploration Geophysicists; 2012

[9] Farquharson CG, Mosher CRW. Three-dimensional modelling of gravity data using finite differences. Journal of Applied Geophysics. 2009;**68**(3):417-422

[10] Jahandari H, Farquharson CG. Forward modeling of gravity data using finite-volume and finite-element methods on unstructured grids. Geophysics. 2013;**78**(3):G69-G80

[11] Jin JM. The Finite Element Method in Electromagnetics. 2nd ed. John Wiley & Sons; New York. 2002

[12] Barbosa VCF, Silva JB, Medeiros WE. Gravity inversion of basement relief using approximate equality constraints on depths. Geophysics. 1997;**62**(6):1745-1757

[13] Barbosa VC, Silva JB, Medeiros WE. Gravity inversion of a discontinuous relief stabilized by weighted smoothness constraints on depth. Geophysics. 1999;**64**(5):1429-1437

[14] Gallardo-Delgado LA, Pérez-Flores MA, Gómez-Treviño E. A versatile algorithm for joint 3D inversion of gravity and magnetic data. Geophysics. 2004;**68**(3):949-959

[15] Martins CM, Barbosa VC, Silva JB. Simultaneous 3D depth-to-basement and density-contrast estimates using gravity data and depth control at few points. Geophysics. 2010;**75**(3):I21-I28

[16] Martins CM, Lima WA, Barbosa VC, Silva JB. Total variation regularization for depth-to-basement estimate: Part 1—Mathematical details and applications. Geophysics. 2011;**76**(1): I1-I12

[17] Silva JB, Costa DC, Barbosa VC. Gravity inversion of basement relief and estimation of density contrast variation with depth. Geophysics. 2006;**71**(5):J51-J58

[18] Silva JB, Oliveira AS, Barbosa VC. Gravity inversion of 2D basement relief using entropic regularization. Geophysics. 2010;**75**(3):I29-I35

[19] Cai H. Advanced methods for depth-to-basement estimation using gravity, magnetic, and electromagnetic data [dissertation]. The University of Utah; Salt Lake City. 2015

[20] Cai H, Zhdanov M. Application of Cauchy-type integrals in developing effective methods for depth-to-basement inversion of gravity and gravity gradiometry data. Geophysics. 2015;**80**(2):G81-G94

[21] Hansen PC, O'Leary DP. The use of the L-curve in the regularization of discrete ill-posed problems. SIAM Journal on Scientific Computing. 1993;**14**(6):1487-1503

[22] Čuma M, Wilson GA, Zhdanov MS. Large-scale 3D inversion of potential field data. Geophysical Prospecting. 2012;**60**(6):1186-1199

[23] Zhdanov MS, Cai H, Wilson GA. 3D inversion of SQUID magnetic tensor data. Journal of Geology & Geosciences. 2012;**1**(104):2

[24] Tarantola A. Inverse problem theory and methods for model parameter estimation. Society for Industrial and Applied Mathematics. 2005

[25] Shamsipour P, Marcotte D, Chouteau M, Keating P. 3D stochastic inversion of gravity data using cokriging and cosimulation. Geophysics. 2010;**75**(1):I1-I10

[26] Shamsipour P, Chouteau M, Marcotte D. 3D stochastic inversion of magnetic data. Journal of Applied Geophysics. 2011;**73**(4):336-347

[27] Portniaguine O, Zhdanov MS. Focusing geophysical inversion images. Geophysics. 1999;**64**(3):874-887

[28] Chakavarthi V, Sundararajan N. Gravity anomalies of 2.5-D multiple prismatic structures with variable density: A Marquardt inversion. Pure and Applied Geophysics. 2006;**163**(1): 229-242

[29] Zhdanov MS, Cox LH. Multinary inversion for tunnel detection. IEEE Geoscience and Remote Sensing Letters. 2013;**10**(5):1100-1103

[30] Krahenbuhl RA, Li Y. Hybrid optimization for lithologic inversion and time-lapse monitoring using a binary formulation. Geophysics. 2009;**74**(6):I55-I65

[31] Zhdanov MS. Inverse Theory and Applications in Geophysics. Elsevier; Amsterdam. 2015

[32] Gallardo LA, Meju MA. Joint two-dimensional DC resistivity and seismic travel time inversion with cross-gradients constraints. Journal of Geophysical Research: Solid Earth. 2004;**109**(B3):1-11

[33] Gallardo LA. Multiple cross-gradient joint inversion for geospectral imaging. Geophysical Research Letters. 2007;**34**(19):1-5

[34] Gallardo LA, Fontes SL, Meju MA, Buonora MP, de Lugao PP. Robust geophysical integration through structure-coupled joint inversion and multispectral fusion of seismic reflection, magnetotelluric, magnetic, and gravity images: Example from Santos Basin, offshore Brazil. Geophysics. 2012;**77**(5):B237-B251

[35] Cai H, Zhdanov MS. Joint inversion of gravity and magnetotelluric data for the depth-to-basement estimation. IEEE Geoscience and Remote Sensing Letters. 2017;**14**(8):1228-1232

Section 2

Gravity in Geoscience Applications

Chapter 3

Microgravity and Its Applications in Geosciences

Hakim Saibi

Additional information is available at the end of the chapter

http://dx.doi.org/10.5772/intechopen.71223

Abstract

Gravity is the most important force which determines the structure and evolution of stars like the Sun as well as the structure and evolution of galaxies. The law of universal gravitation is generally sufficient to describe the gravity of the Earth, the Moon, or the planets orbiting the Sun. With the recent development of sensitive gravimeters, the gravity survey has become one of the most used geophysical tools in applied geosciences for tasks including: exploring for oil and gas fields by studying geological structures and salt dome intrusion, monitoring groundwater and geothermal reservoirs by determining recharge and discharge masses, monitoring volcanic activity and hydrothermal activity beneath volcanoes, monitoring CO_2 movement during and after sequestration, locating active faults responsible for big earthquakes, and also exploring mines and detecting local cavities. In this chapter, we present a brief introduction to gravity and Bouguer gravity, the different corrections applied to measured gravity and follow with cases of applied microgravity measurements in different fields of geosciences.

Keywords: gravity, faults, geological structures, volcanoes, groundwater, geoscience

1. Introduction

The gravity method is a nondestructive geophysical technique that measures differences in the Earth's gravitational field between specific locations. It has many applications in engineering and environmental studies such as locating karsts, monitoring aquifer recharge, determining geologic layer thickness and the structure of the basement rocks, estimating the mass and volume changes in geothermal reservoirs, monitoring precursors of volcanic eruptions, and monitoring gas production and carbon sequestration. The gravity method is also used in oil, gas, and mineral exploration.

The gravity method depends mainly on the differences in the density of the Earth materials. The variations of densities of subsurface rocks produce variations in the measured gravity field. There are many numerical and analytical methods to study the variations in gravity and interpret the source of these variations (geometry, depth, and density). A GPS measurement should be associated with gravity measurements in order to know the exact coordinates (longitude and latitude) of the gravity stations and their altitudes. The measured gravity data is then processed by removing all the quantifiable disturbing effects and interpreted using computer programs. The most highly processed data are known as the Bouguer gravity, and anomalies are measured in units of mGal. Recent computer models are capable of creating three-dimensional subsurface density models.

2. Gravity method

2.1. Definition of the gravity field

The gravity field is defined as follows: a gravitational force or gravity exerted on a unit mass at a point in space or on the surface of the earth or its vicinity [1]. The gravity field is a force or a force field distributed over the region surrounding the generator. A measurement of gravity fields should be made in a region or space in which gravity fields exist.

2.2. Gravity method and its applications

Gravity is a geophysical potential field method. Various applications and developments of gravity method in the fields of geology, engineering, geothermal, and volcanology have been demonstrated and developed. Gravity can be used in time-series applications such as geothermal reservoirs and by estimating the underground mass changes [2, 3]. Also, gravimetric studies at active volcanic systems have contributed significantly to the better understanding of pre-, syn-, and post-eruptive processes [4, 5]. With the development of gravimeters with improved accuracy, the measurements have become much easier. Recently, time-variable gravity from satellites has detected the redistribution of mass over large scales and the gravity data are used, for example, in hydrology and oceanography. The gravity method can also detect geological anomalies and contacts (faults) by using gravity gradient interpretation techniques.

2.3. Basics of gravity

Gravity surveying is based on Isaac Newton's universal law of gravitation, described in *Principia Mathematica* in 1687. One of the basic forces of nature is the attraction between all masses. This attraction is called the force of gravity. According to the Newtonian law of gravity, the gravitational force between any two point masses is given by:

$$F = \frac{G\ M_1\ M_2}{r^2} = gm \tag{1}$$

where G (universal gravitational constant) = 6.673×10^{-11} m^3 kg^{-1} s^{-2}. M_1 and M_2 are the two masses in kg, and r is the distance between the point masses, in meters. g is the local value of the Earth's field, and m is a test mass. This force acts in the direction joining the two masses (**Figure 1**). **Figure 2** shows the gravitational attraction of a small mass on the Earth.

2.3.1. Units of gravity

Recent gravity meters are very sensitive and can routinely measure differences in the gravity field to within one part in 10^9. The c.g.s unit commonly used in gravity measurement is the milliGal: 1 mGal = 10^{-3} Gal = 10^{-3} cm s^{-2}. In gravity surveys, commonly, mGal is used.

2.3.2. Shape of the Earth

If the Earth was a uniform, homogenous sphere, g would be constant over its surface. However, gravity varies because the density varies within the Earth, and the Earth is not a perfect sphere. One predictable effect on local gravity measurements is earth's shape. The Earth has the shape of a flattened sphere because of its rotation (**Figures 3–5**).

2.4. The pendulum

It is possible to find the acceleration of gravity (g) from the period of oscillation of a pendulum (**Figure 6**) swinging with a small amplitude using Eq. (2):

$$g = (2\pi f)^2 L \tag{2}$$

where f is the frequency and L is the length.

2.5. Modern gravimeters

There are two main types of modern gravimeters. The first type, which uses a pendulum arrangement or a dropping weight, records the actual acceleration of gravity wherever it is placed. The second type uses levers and springs to measure the difference in gravity between two stations. Since the springs are not calibrated, these readings are all relative to a base station.

Gravimeters, essentially a mass suspended from a sophisticated spring balance, have been used to measure relative gravity since 1930s. As weight of mass (mass x gravity) increases, the spring is stretched (**Figure 7**).

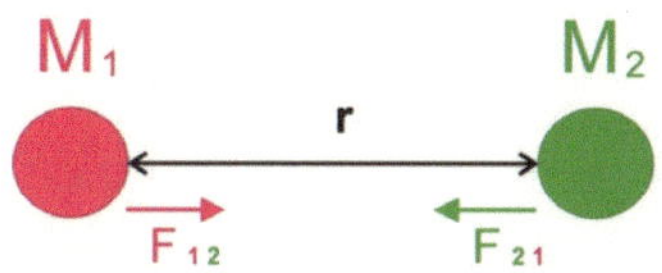

Figure 1. Attractive force acting between two bodies.

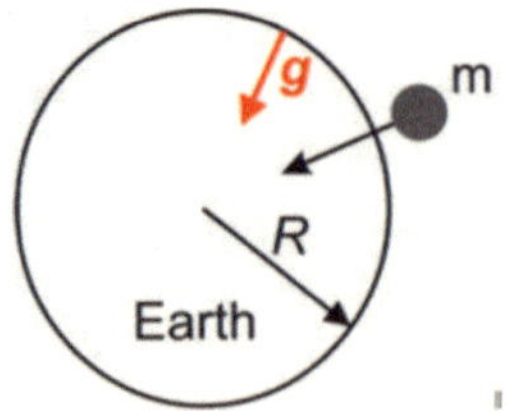

Figure 2. The gravitational attraction of a small mass m on the Earth.

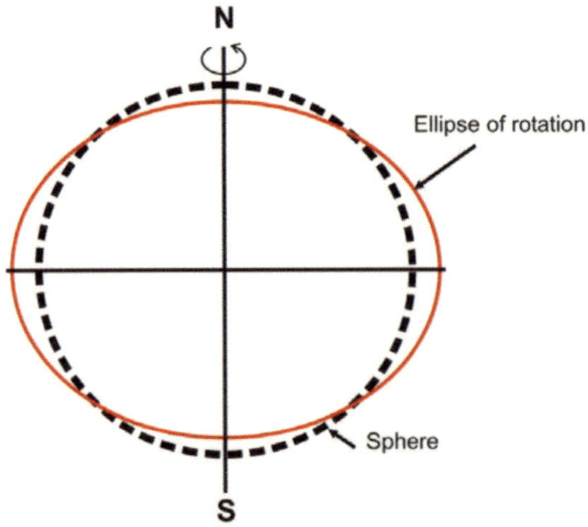

Figure 3. Exaggerated difference between a sphere and an ellipse of rotation (spheroid).

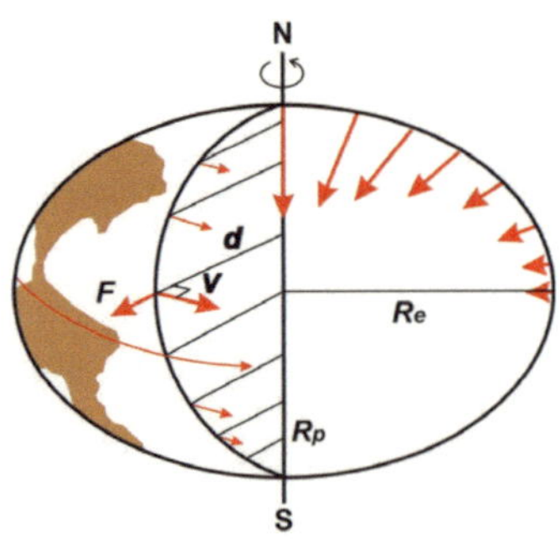

Figure 4. Centrifugal acceleration and variation of gravity with latitude ϕ (not to scale). The gravity is 5186 mGal greater at the poles than at the equator. The acceleration due to gravity varies with latitude due to two effects: The Earth's shape and the Earth's rotation [6].

2.5.1. Relative modern gravimeter

2.5.1.1. CG-3M relative gravimeter

The Scintrex CG-3M gravimeter is a very sensitive mechanical balance that detects gravity field changes as small as one part in a million. CG-3M (**Figure 8**) has a resolution of 1 μGal

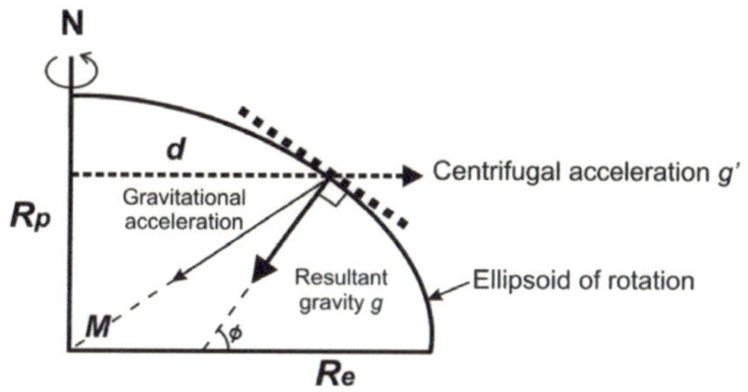

Figure 5. Resultant of centrifugal acceleration (g') and the acceleration due to gravity (g) [6].

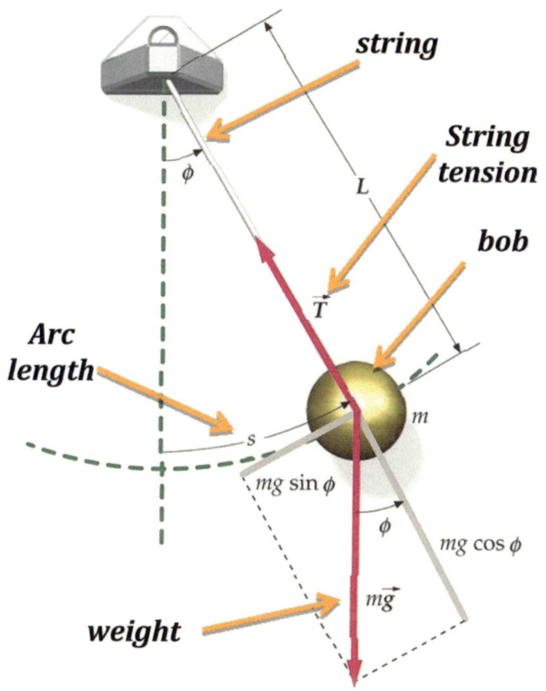

Figure 6. The pendulum.

and automated corrections for tide, instrument tilt, temperature, and rejection of noisy data. **Figures 9** and **10** show the CG-3M gravimeter in the field with GPS measurement. Please check **Table A1** for gravity data sheet information.

2.5.1.2. CG-6 relative gravimeter

The CG-6 Autograv (Scintrex) is the newest generation of land gravity meter. **Figure 11** shows CG-6 acquired by United Arab Emirates University in March 2017. **Figure 12** shows the gravity stations observed in Al Ain city. **Figure 13** shows the three-dimensional inversion of the Bouguer gravity data for geological investigations.

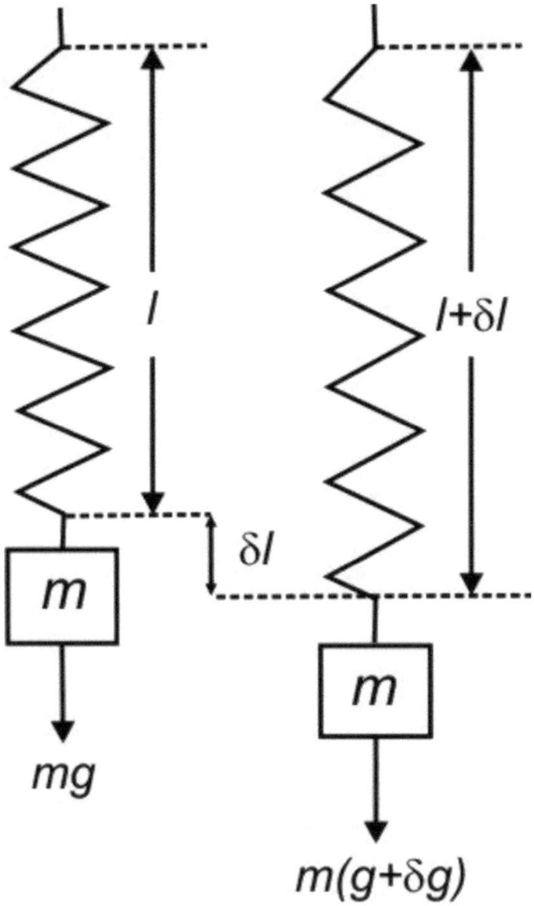

Figure 7. Extension (δl) of a spring due to additional gravitational pull (δg).

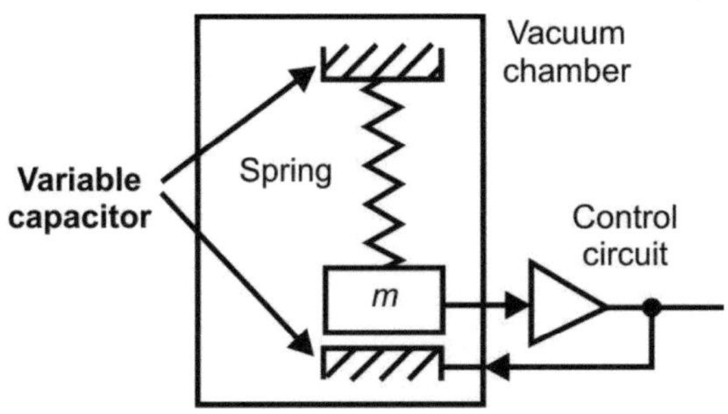

Figure 8. Scintrex CG-3 gravimeter.

2.5.2. Absolute modern gravimeter

The A-10 (**Figure 14**) is a portable absolute gravimeter recently developed by Micro-g LaCoste, Inc. (MGL) that is designed for use in the field to measure the vertical acceleration of gravity (*g*).

Figure 9. Photograph showing the Scintrex CG-3M gravimeter and GPS antenna at Nita station (Unzen volcano, southwestern Japan) [7].

Figure 10. Photograph of CG-3M Autograv automated gravity meter. The features of this gravimeter are: resolution mGal = 0.001 (=1 μGal), standard deviation: <5 μGal, automated corrections: tide, instrument tilt, temperature, and rejection [7].

Figure 11. Top view of CG-6 gravimeter during field measurement in Alain city, United Arab Emirates (June 2017) [7].

Figure 12. View of lynx LG software (Scintrex, 2017) showing the gravity stations in Alain city, United Arab Emirates, during field survey in Al-Ain city in June 2017. Lynx LG software includes data pre- and postprocessing that helps users to get a simple Bouguer gravity map of the study area just after finishing the measurements [7].

A test mass is dropped numerous times in a vacuum, and its position is measured with a laser interferometer as a function of time with an atomic clock. The vertical acceleration of gravity is calculated by fitting the equation of motion to the measured trajectory of the test mass. The drops are combined into a "set" which typically consists of 100–200 drops. Multiple sets are

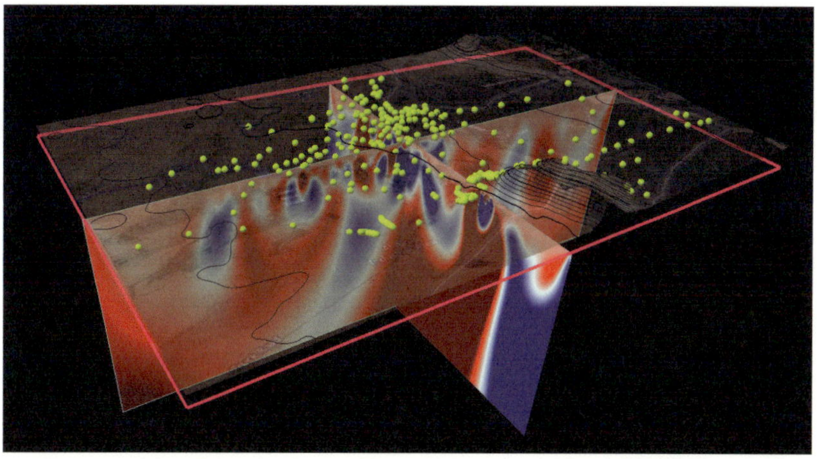

Figure 13. Three-dimensional gravity inversion of Alain city Bouguer gravity data using petrel gravity magnetic modeling and inversion plug-in (Schlumberger) showing regions of high and low densities beneath the study area [7].

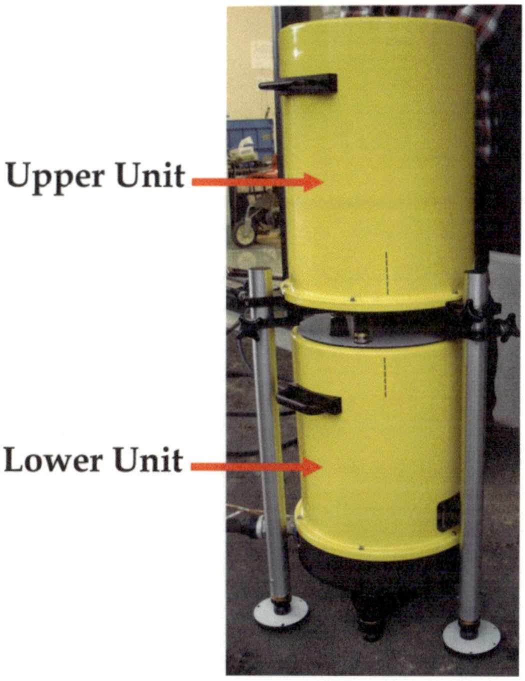

Figure 14. Picture of A-10 absolute gravimeter.

collected, and the average of the sets provides a value of *g*. The specifications of the A-10 are: a precision of 100 μGal at a quiet site, a repeatability of 10 μGal on a high-quality pier, and an accuracy of 10 μGal [8]. **Table 1** and **Figure 15** show the absolute gravity data recorded in Kyushu University [7]. Please check **Table B1** for gravity data sheet information.

Setup height	2.20 cm
Transfer height	100.00 cm
Actual height	74.00 cm
Gradient	−3.086 μGal/cm
Nominal air pressure	1004.27 mBar
Polar motion coord.**	−0.1049 "0.2862"
Measurement precision	0.94 μGal
Number of sets	10
Number of drops	100
Gravity	979634789.69 μGal

**The polar motion changes daily. From the MGL website, we can download the values of polar motion and enter them in the computer before processing the gravity data.

Table 1. Example of absolute gravity measurements using A10 absolute gravimeter [7].

Measurement of absolute gravity at base station is very important in calculating absolute gravity values at other stations as follows:

$$G_{abs} = G_{absB} + G_{obs} - G_B + G_H + G_T + G_D$$

G_{abs} : absolute gravity value at the station
G_{absB} : absolute gravity value at the base station
G_{obs} : measured value at the station(mgal converted)
G_B : measured value at the base station
G_H : instrument height correction value at the station
G_T : tide correction value at the station
G_D : drift correction value at the station

2.6. Corrections to gravity observations

2.6.1. Instrumental drift

Gravimeters are very sensitive instruments. Temperature changes and elastic creep in springs cause meter readings to change gradually with time even if the meter is never moved. Drift is monitored by taking repeated readings at the same station over the course of the day, perhaps every 1–2 h, to produce a drift curve (**Figure 16**). Instrument drift correction for each station can be estimated from drift curve.

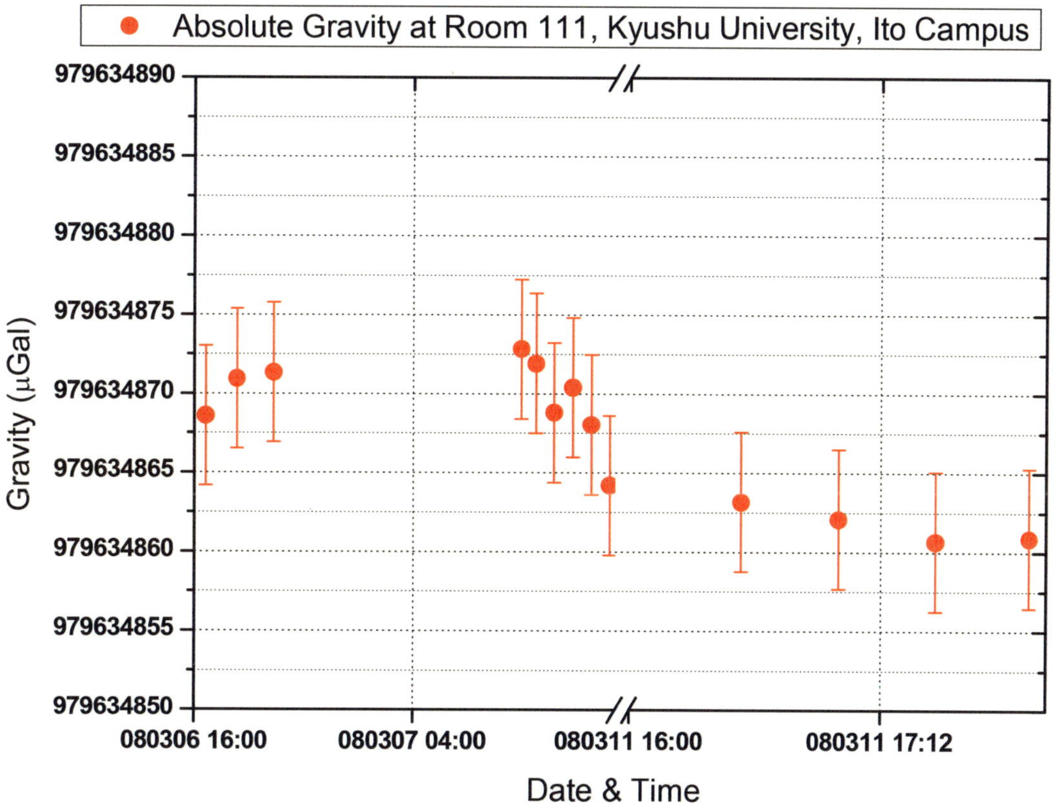

Figure 15. Change in absolute gravity measured by the absolute gravimeter A-10 at Kyushu University (Japan) after applying tidal, polar motion, and air pressure corrections. Measurement precision is 0.94 μGal; the number of sets is 10; the number of drops is 100. Analysis of the data indicates the instrument performed within the specifications of the manufacturer. The microgravity changes are associated with the shallow groundwater change due to rainfall [7].

2.6.2. Variation with elevation

2.6.2.1. Free-air effect

Height correction is important in a microgravity survey. The measured gravity is corrected by using free-air gradient of −308.6 μGal/m (**Figure 17**).

Free-air correction is the difference between gravity measured at sea level and at an elevation, h, as if there was no rock in between.

2.6.2.2. Bouguer effect

Free-air correction does not take into account the mass of rock between measurement station and sea level. The Bouguer correction, Δg_B, accounts for effect of the rock mass by calculating extra gravitational pull exerted by rock slab of thickness *h* and mean density ρ (**Figure 18**).

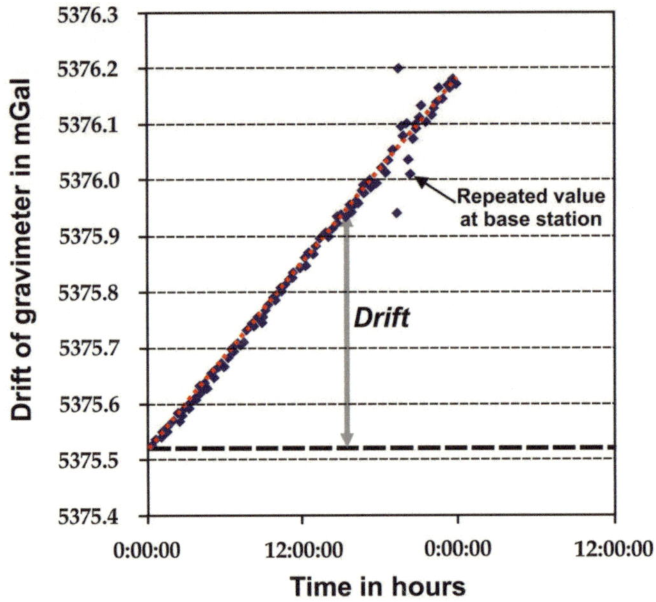

Figure 16. An example of instrumental drift correction [7].

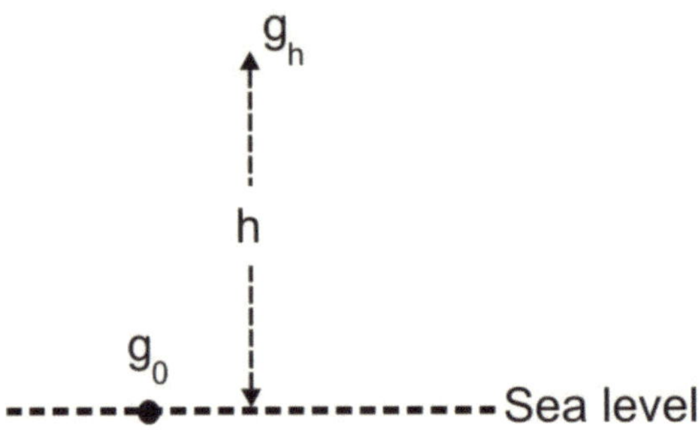

Figure 17. Free-air effect.

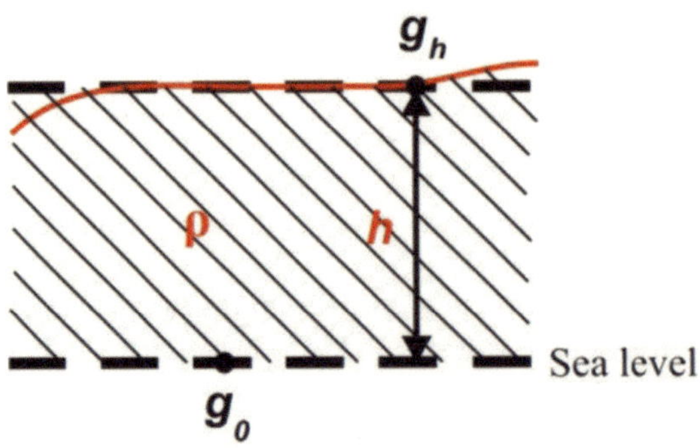

Figure 18. Bouguer effect.

$$\Delta g_B = +0.04192\ h\ \rho\,\text{mgal} \tag{3}$$

2.6.2.3. Elevation effect

$$\Delta g_E = -(0.3086 - 0.0149\rho)\ \text{mgal/m} \tag{4}$$

2.6.2.4. Terrain effect

Bouguer correction assumes subdued topography. Additional terrain corrections must be applied where measurements are near to mountains or valleys (**Figures 19–21**). If the station is next to mountain, there is an upward force on the gravimeter from mountain that reduces the reading. This effect was first noticed in topographic surveys run near the Himalayan Mountains.

If gravity station is next to valley, there is an absence of the downward force on gravimeter assumed in Bouguer correction, which reduces free-air anomaly.

In both cases, terrain correction is added to Bouguer anomaly.

2.6.3. Variation with time

2.6.3.1. Earth and ocean tides

The solid Earth responds to the pull of the Sun and the Moon just like the oceans, but movements are much smaller. The Sun and the Moon also pull on the gravimeter and its parts. These effects are large enough to affect gravity readings (**Figure 22**). Changes in observed

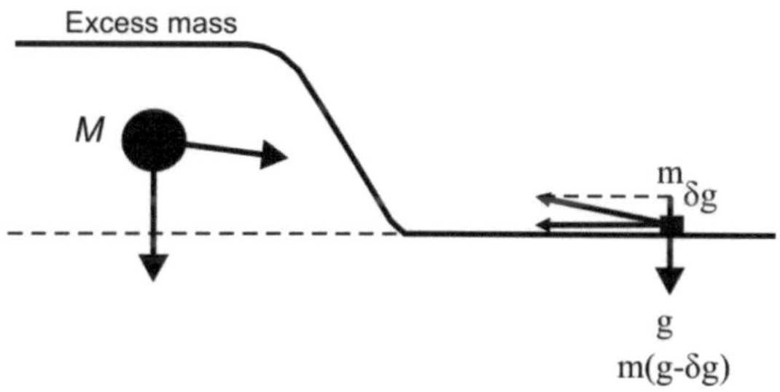

Figure 19. Terrain effect.

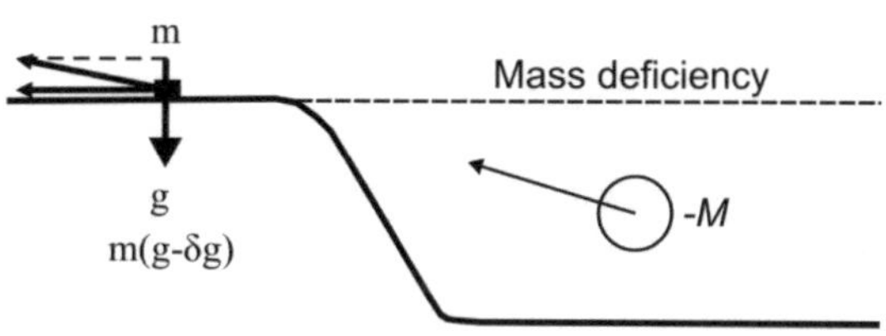

Figure 20. Terrain effect.

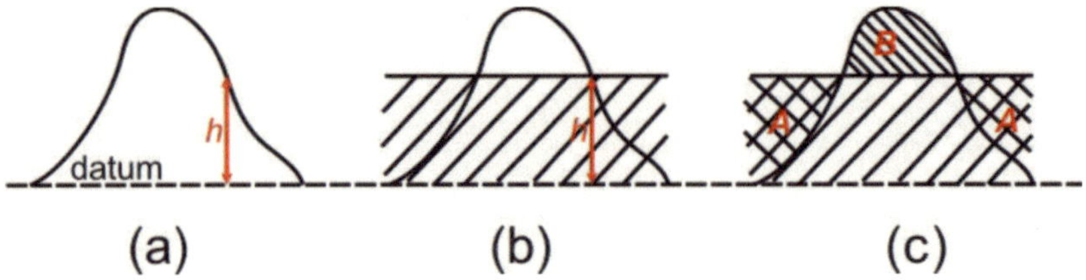

Figure 21. Examples of terrain correction. (a) Gravity station at elevation *h* above geoid. (b) Infinite rock mass assumed for Bouguer correction. (c) Terrain correction compensates for error *A* in *BC* and also *B*.

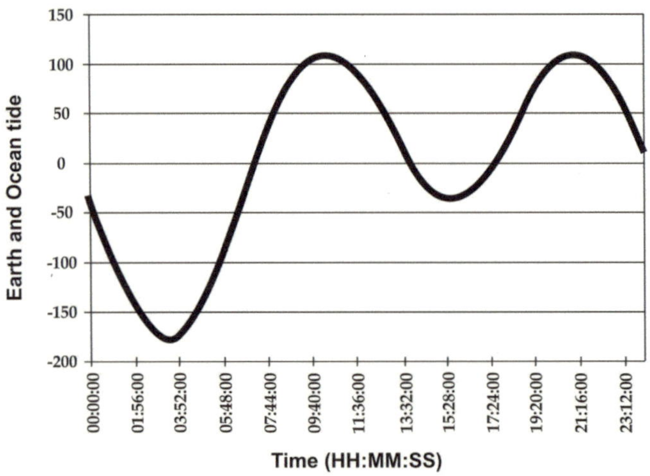

Figure 22. An example of earth tide and ocean load tide effects in μGal at a gravity station in Unzen volcano (southwestern Japan) on August 11th, 1999 using GOTIC2 computer code [7, 9].

gravity due to "tides" occur with periods of 12 h or so. Earth tide effects can be controlled by repeated readings at same station in same way as instrument drift.

Precipitation and atmospheric pressure: These effects can be significant in a microgravity survey. It is recommended to measure gravity in a stable weather condition (on the same day, for example).

2.7. Bouguer anomaly

The Bouguer anomaly (Dg_B) is the difference between the observed value (g_{obs}), properly corrected, and a value at a given base station (g_{base}), such that:

$$\Delta g_B = g_{\mathrm{base}} + \sum(\mathrm{corr}) - g_{\mathrm{obs}} \tag{5}$$

with

$$\sum(\mathrm{corr}) = \delta g_L + \left(\delta g_F - \delta g_B\right) + \delta g_{\mathrm{TC}} - \delta g_D \tag{6}$$

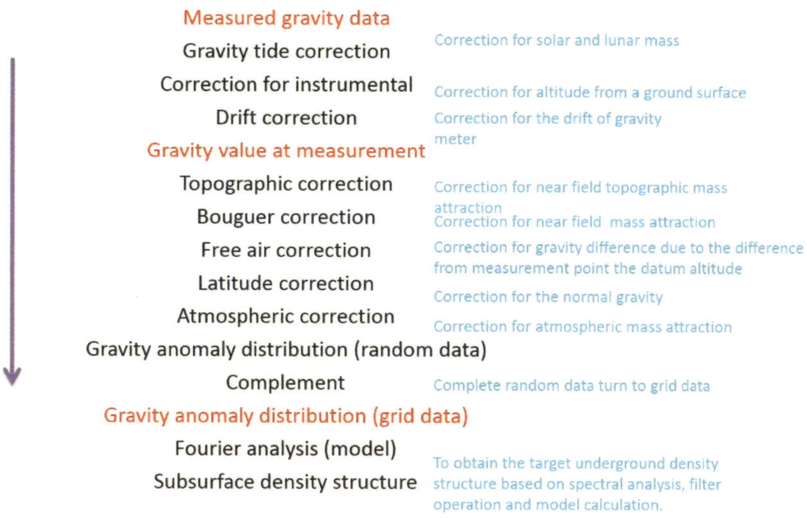

Figure 23. Flow chart of gravity data processing and analysis.

where the subscripts refer to the following corrections: L: latitude; F: free-air; B: Bouguer; TC: terrain correction; and D: drift (including the Earth tides).

Figure 23 shows the different steps for gravity data processing.

3. Applications of gravity method in geosciences with case studies

Gravity can be used for: (1) regional surveys and (2) local surveys. These two cases depend on the scale of the geologic target under the study. For fault and regional geological surveys, we use a regional gravity survey with a large grid size. However, for local geological studies such as cavity investigations, active fault detection or local geological features, we use more dense gravity surveys with a small grid size and small spacing between gravity stations.

3.1. Geology and structural geology

Gravity data can give us much information about the subsurface geological structures such as faults, rock intrusions, dykes, and sills. We can also study basin structure, grabens, horts, and salt intrusion all of which are very important in oil and gas exploration. The most important physical property is rock density, and density contrast between the different rocks will create gravity anomalies which are easy to detect if the measured gravity data are good enough (**Figure 24**). Gravity measurements are very sensitive to noises such as cars and human activities including walking vibrations from machinery or nearby seismic operations and even earthquakes. The geophysicist and/or user must pay close attention during field measurement in order to get good gravity data.

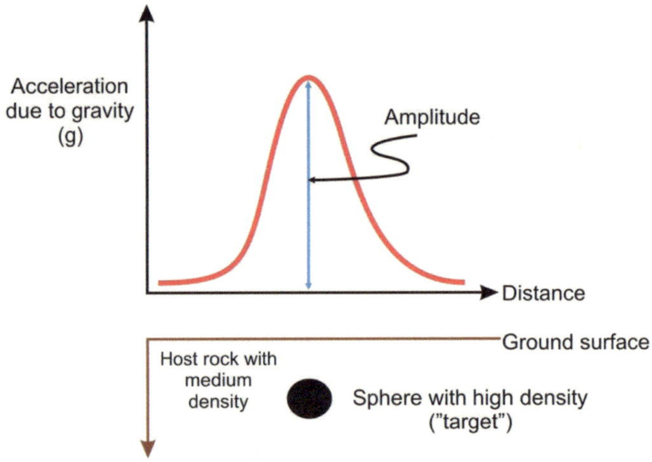

Figure 24. Example of gravity anomaly over a buried sphere with a higher density than the surrounding rocks.

Bouguer anomaly maps contain both regional and residual (local) anomalies (**Figure 25**). Regional anomalies have long wavelengths and are usually due to deep crustal features. A residual (local) anomaly will have a short wavelength and is due to shallower structures.

There are many filtering techniques that can help us to separate the residual and the regional gravity anomalies such as band-pass filters. Power spectrum analysis techniques are also useful in determining the number of geological layers and their depths and can also be used to remove noise from the data.

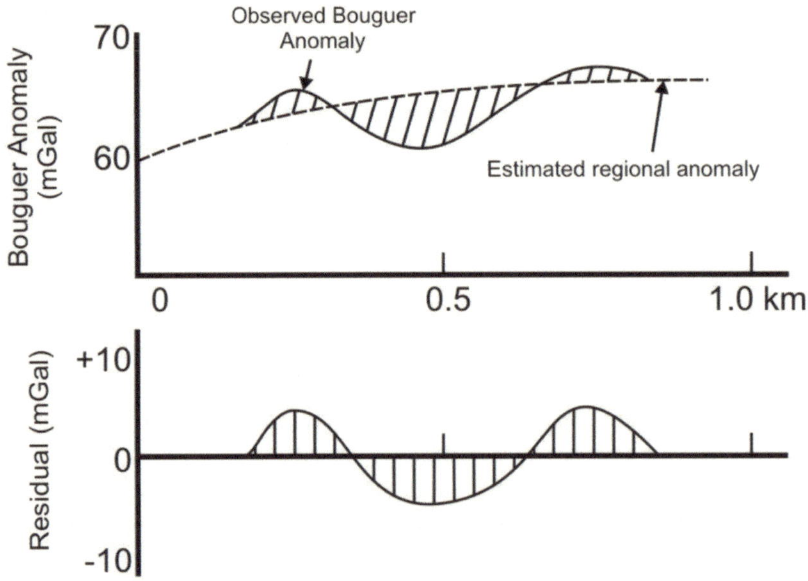

Figure 25. Removal of a residual gravity from a regional profile.

3.1.1. Forward modeling

Figure 26 shows an example of forward modeling of gravity data from Aynak-Logar Valley (Afghanistan) for studying the local geology.

3.1.2. Three-dimensional gravity inversion, case study: Tindouf Basin (Algeria)

The gravity effect of the three-dimensional (3D) density model for Tindouf basin has been computed using GRABLOX-1.7 and BLOXER-1.5 software developed by M. Pirttijärvi, University of Oulu, Finland [11].

The model of study area covers an area of 42,501 km^2 and was oriented in a north–south direction, extending 269 km in the east–west direction and 157 km in a north–south direction (**Figure 27**). The spatial discretization was 83 grid blocks in the east–west direction ("i" index)

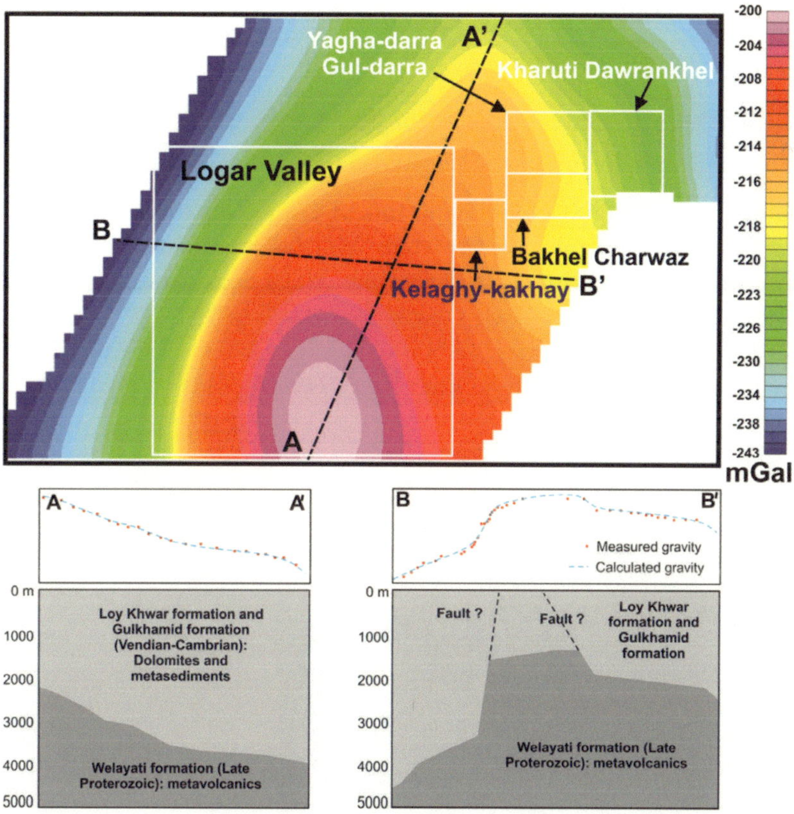

Figure 26. Example of utilization of Bouguer gravity data to construct geological models. Top figure shows Bouguer gravity map at Aynak-Logar valley in Afghanistan with a high gravity anomaly located in southern part of the study area. The figures below show two-dimensional (2D) forward modeling of gravity data using [10] computer code after assigning density values for Loy Khwar and Welayati formations. We can observe a good matching between observed and calculated gravity values. To check the goodness of the geological model, we need to add well data (depth of boundary between the two formations) and use it as control point when creating 2D forward models. [7].

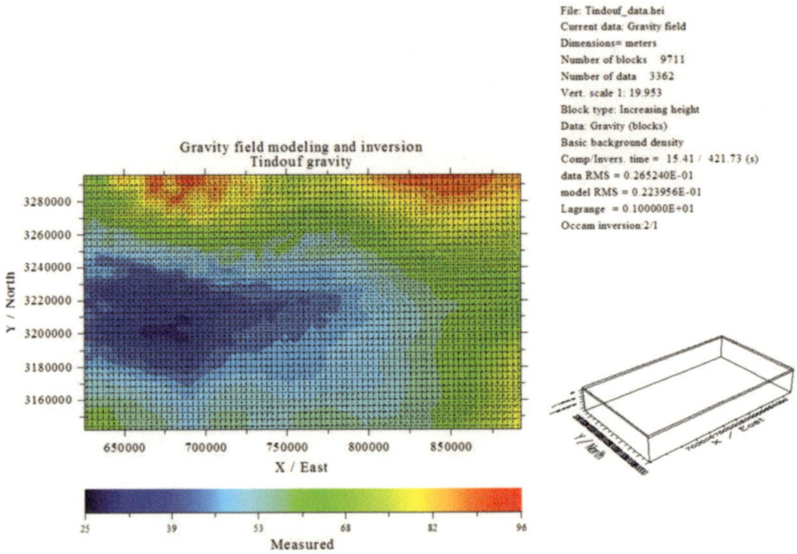

Figure 27. Measured Bouguer airborne gravity data of Tindouf basin and the mesh characteristics in preparation for the 3D inversion [7].

and 39 blocks in the north–south direction ("j" index). In the Z-direction ("k" index), the model contains three blocks. The model is based on two-layer case: sedimentary layer represented by clays with a density of 2.2 g/cm^3 and a dolerite layer with a density of 2.9 g/cm^3. **Figures 28–32** show the 3D gravity inversion results.

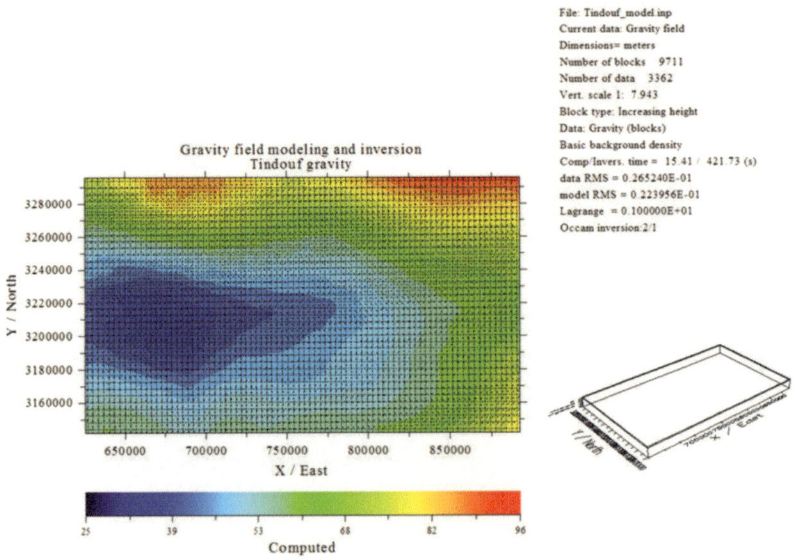

Figure 28. Computed Bouguer gravity of Tindouf basin [7].

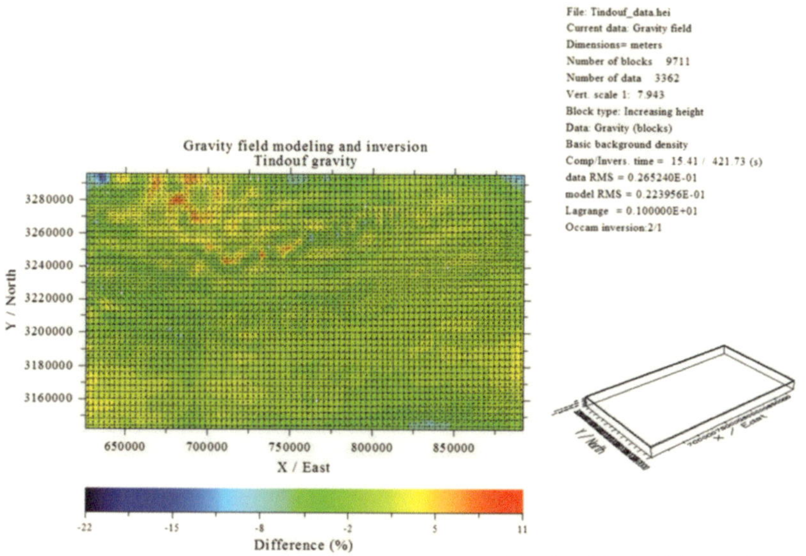

Figure 29. Difference between the computed and the observed Bouguer gravity. Generally, the difference is ranging from 1 to 10% [7].

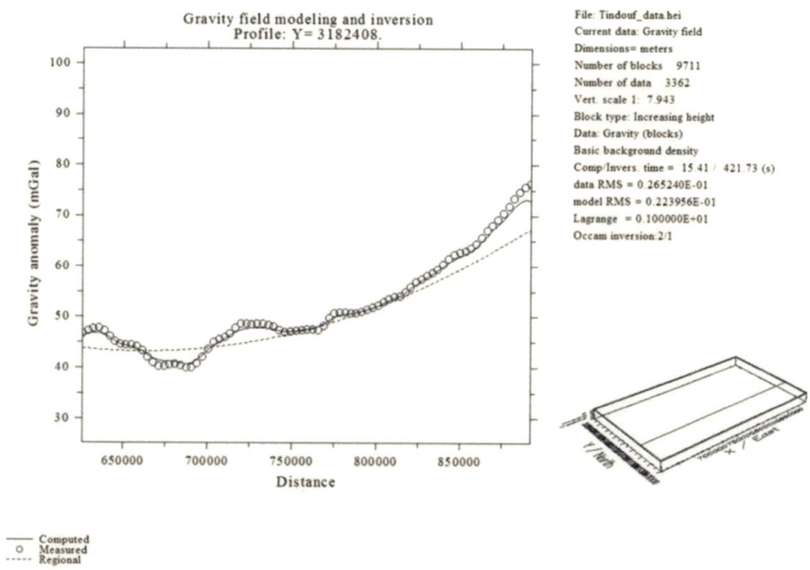

Figure 30. Comparison plot showing the computed, measured, and regional Bouguer gravity at E-W profile [7].

3.2. Geothermal

Time-lapse gravity surveys at geothermal fields have been going on for at least four decades. The movement of fluids and mass redistribution in geothermal reservoirs are essential for geothermal energy development and its sustainability (**Figure 33**). [12] explained many

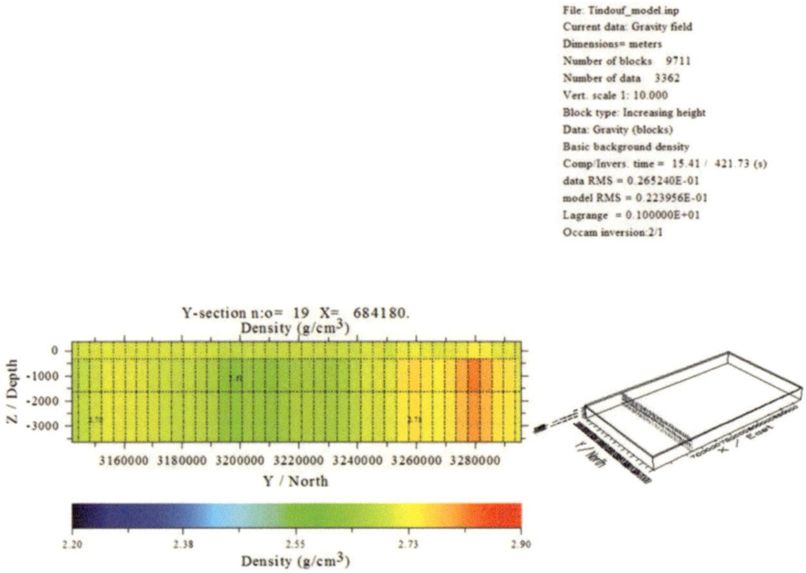

Figure 31. Section of gravity inversion results at an N-S profile showing high-density rock intrusion (dolerite) at the southern part of the profile [7].

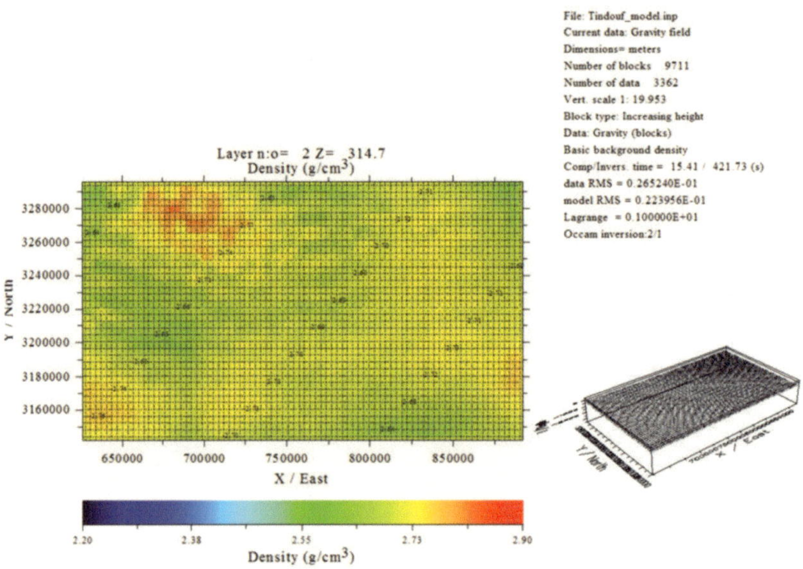

Figure 32. Top view of gravity inversion results at a depth of 314.7 m (Z is positive downward) showing high-density values in the northwestern part of the study area [7].

possible causes of gravity change at Wairakei geothermal field. Recently, geothermal reservoirs have been monitored with a high-precision hybrid technique (absolute/relative gravity-measurement) [13, 14]. **Figure 34** shows the gravity changes in Obama geothermal field (southwestern Japan).

http://dx.doi.org/10.5772/intechopen.71223

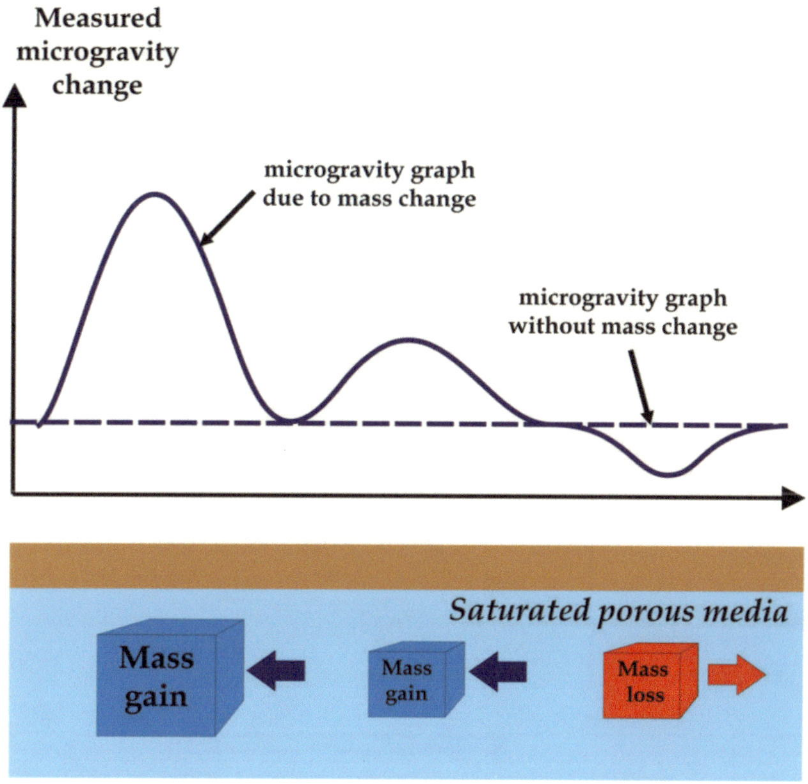

Figure 33. A figure explaining the relation between microgravity changes and mass changes in a geothermal reservoir due to the production and reinjection of waters.

3.3. Hydrogeology

Hydrogeology is perhaps the most complex use of intermediate scale (hour–year) variations in gravity [15]. In the case of horizontal layer (Bouguer slab) of aquifer of thickness h and porosity ϕ, we get a gravity perturbation [16],

$$\Delta g = 2\pi G \rho_{water} \phi h \tag{7}$$

The groundwater level change is calculated after rearranging Eq. (7) as follows:

$$h = \frac{\Delta g}{2\pi G \rho_{water} \phi} \tag{8}$$

Where Δg: gravity change (μGal = 10^{-8} m/s^2), G: universal gravitational constant = 6.673 × 10^{-11} m^3 kg^{-1} s^{-2}, ρ_{water}: density of groundwater, ϕ: porosity of aquifer, and h: water-level change (m). We assume that the groundwater level variations stem from a shallow aquifer (**Figure 35**); however, there is no bound on aquifer depth in this approximation [5].

Figure 36 shows the evaluation of the required groundwater level changes at the shallow subsurface of Unzen volcano (southwestern Japan) calculated using Eq. (8). The maximum values of water-level change are in the range of several meters.

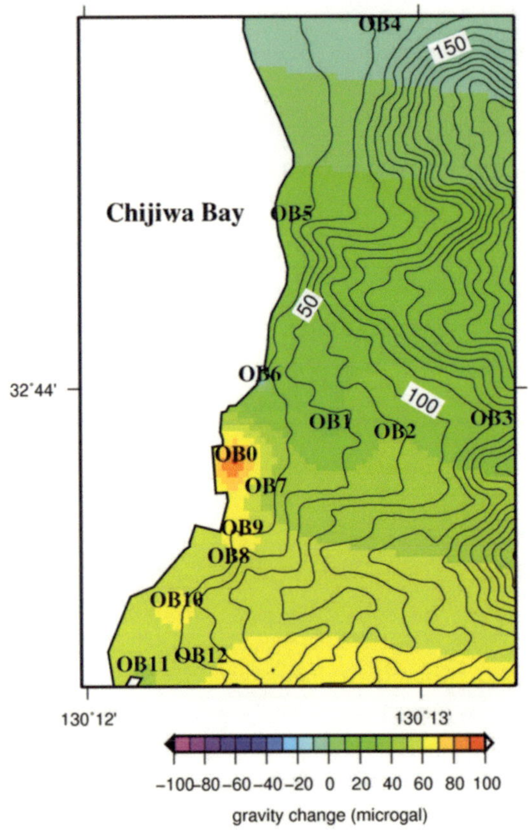

Figure 34. Microgravity changes over the Obama geothermal field (southwestern Japan) from 2003 to 2004. The positive increase of microgravity is interpreted as an excess of mass in the geothermal reservoir [3].

Figure 35. Thickness of a groundwater not deep aquifer.

3.4. Active faults

Gravity method can detect fault zones especially if there is vertical displacement and there is density contrast between the geological layers as shown in **Figure 37**.

3.5. Volcanology

Gravity surveys are very important in monitoring volcanic activity and studying hydrothermal activity beneath volcanoes [5]. **Figure 38** shows the two-may measurements of microgravity in volcanic regions. **Figure 39** explains the different corrections necessary to calculate the residual gravity.

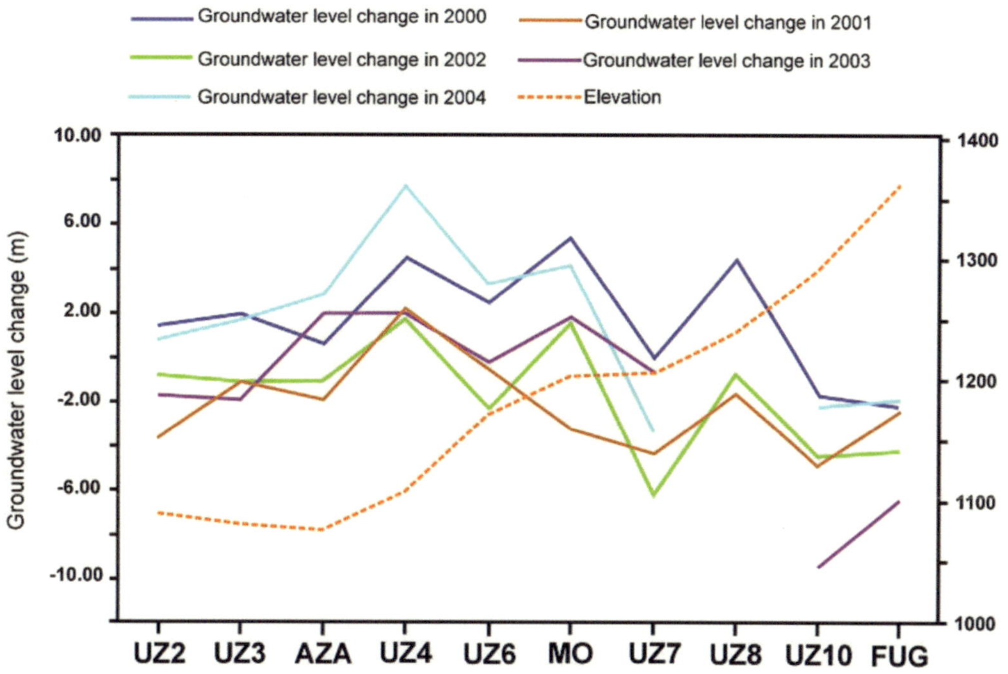

Figure 36. Change of groundwater level with 20% of porosity and topographic elevation along the path crossing stations UZ2 to Fugen microgravity stations in Unzen volcano (Nagasaki prefecture, southwestern Japan). The right vertical axis represents the elevation in meters. The left vertical axis represents the required groundwater level changes in meters [5].

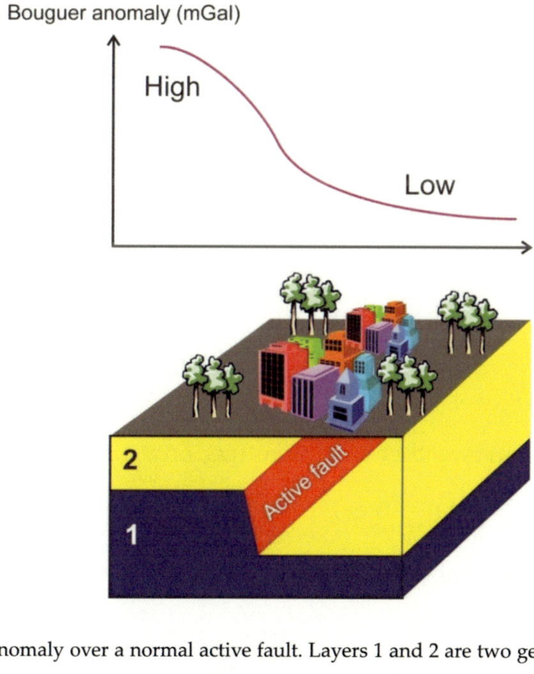

Figure 37. Bouguer gravity anomaly over a normal active fault. Layers 1 and 2 are two geological layers. Layer 1 has a higher density than layer 2.

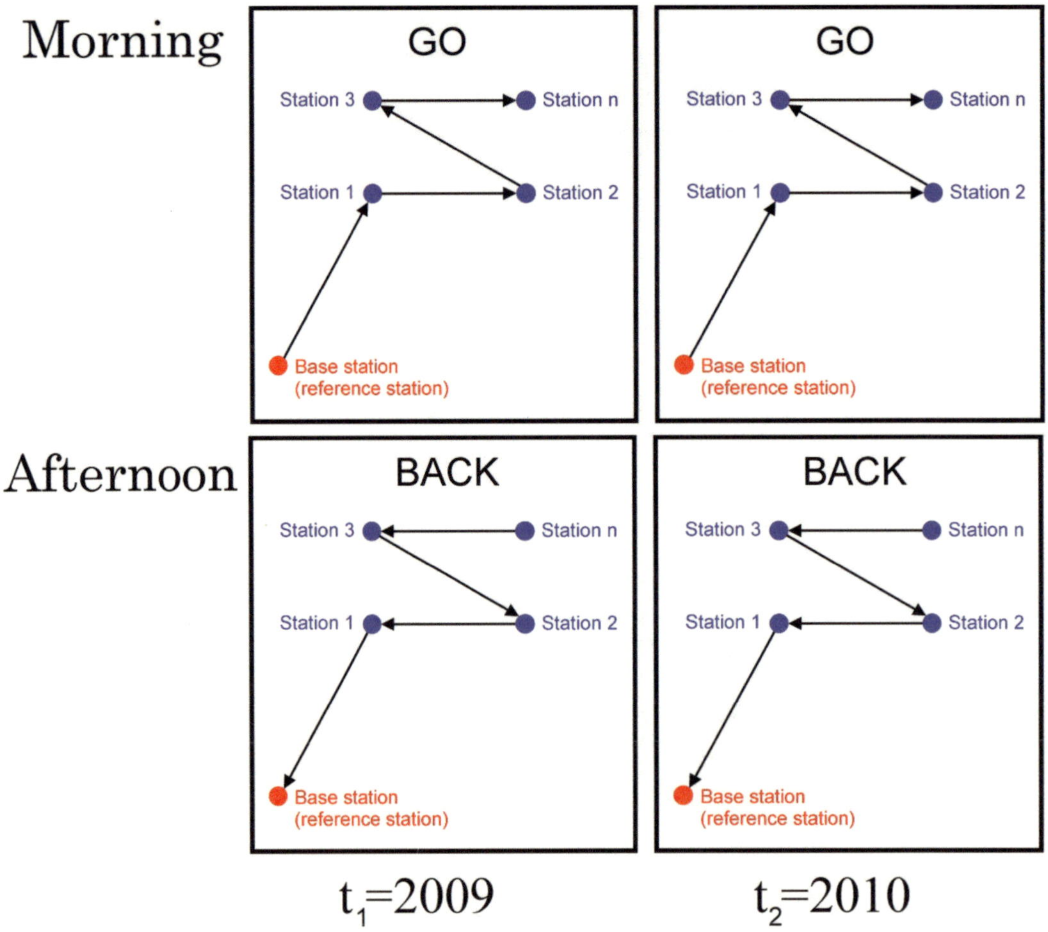

Figure 38. Example of a microgravity monitoring survey at a volcano at different times to monitor volcanic activity.

For gravity surveys in volcanic areas, it is recommended to:

- Use gravity residuals for plotting the gravity changes at a volcano/geothermal field.
- Add the 1-σ error residuals in the same plot of the gravity residuals.
- Use Student's *t*-test to derive indications of statistical relevance of the residual data (to check significant changes at stations).
- Measure gravity in stable weather conditions (i.e., on the same day).
- Minimize operator errors as much as possible by taking the readings in a good way.

3.5.1. Residual gravity calculation

The Residual Gravity at *Station 1* (t_1-t_2) = [(measured gravity at *Station 1* t_2: measured gravity at *Base Station* t_2): (measured gravity at *Station 1* t_1: measured gravity at *Base Station* t_1)].

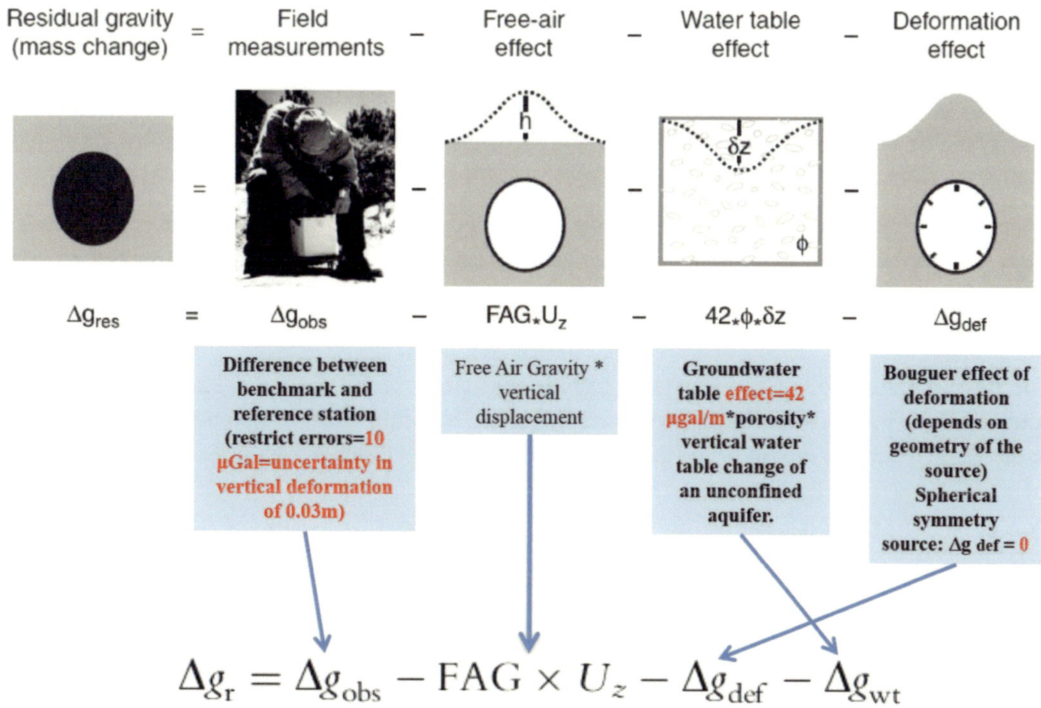

Figure 39. It explains the residual gravity due to residual mass changes at volcanic field [4, 17].

3.5.2. Error datum (1−σ error) calculation

The error for the datum t_1 is as follows:

Error *Station 1* t_1 = sqrt [(error at *Station 1* $t_1)^2$ + (error at *Base Station* $t_1)^2$]

For the datum t_2:

Error *Station 1* t_2 = sqrt [(error at *Station 1* $t_2)^2$ + (error at *Base Station* $t_2)^2$].

error at Station = sqrt [(drift corrected gravity)go: (drift corrected gravity)back]^2.

3.5.3. Calculation of errors in case of ground deformation

The error of the residual gravity changes (1−σ error residuals) is calculated via:

Residual error *Station 1* t_2 = sqrt [(error *Station 1* $t_1)^2$ + (error *Station 1* $t_2)^2$ + (X μGal$)^2$].

Residual error *Station 1* t_3 = sqrt [(error *Station 1* $t_1)^2$ + (error *Station 1* $t_3)^2$ + (2*X μGal$)^2$].

Residual error *Station 1* t_n = sqrt [(error *Station 1* $t_1)^2$ + (error *Station 1* $t_n)^2$ + ((*n*−*1*)*X μGal$)^2$].

X μGal is Propagated Error Uncertainty (i.e., deformation effect).

4. Interpretation techniques

In general, we have three categories of interpretation techniques of gravity data: (1) gradient interpretation techniques based on the derivatives of the gravity field, (2) forward modeling of gravity data, and (3) inversion techniques of gravity data in 2D and 3D.

There are many gravity gradient methods developed for detecting fault structures, intrusive high-density bodies, and geological boundaries. Each technique has some advantages for specific geological investigations. In **Table 2**, we summarized these methods, which are very powerful for investigating geological structures as applied to real gravity field data [18–22].

For 2D and 3D modeling and inversion techniques using gravity data, there are many available computer codes and also computer software developed by commercial companies. For more details about these inversion methods and computer codes, you can check these publications [23–26].

Method	Equation	Reference	Advantages
Analytic signal	$AS(x,y) = \sqrt{\left(\frac{\partial P}{\partial x}\right)^2 + \left(\frac{\partial P}{\partial y}\right)^2 + \left(\frac{\partial P}{\partial z}\right)^2}$	[27]	
Tilt derivative	$TDR = \tan^{-1}\left\{\frac{\frac{\partial P}{\partial z}}{\sqrt{\left(\frac{\partial P}{\partial x}\right)^2 + \left(\frac{\partial P}{\partial y}\right)^2}}\right\}.$	[28]	Useful in enhancing and sharpening the potential field anomalies. The zero contour line located on or close to a contact.
Horizontal gradient	$\mathrm{HG} = \sqrt{\left(\frac{\partial g}{\partial x}\right)^2 + \left(\frac{\partial g}{\partial y}\right)^2}$	[29]	Least susceptible to noise in the data because it requires only the calculation of the two first-order horizontal derivatives of the field. The method is also robust in delineating both shallow and deep sources.
Euler deconvolution	$(x - x_o)\frac{\partial M}{\partial x} + (y - y_o)\frac{\partial M}{\partial y} + (z - z_o)\frac{\partial M}{\partial z} = n(\beta - M)$	[30]	Assign structural index (0 for fault, 1 for contact, and 2 for a sphere) Help us in detecting linear fault structures and their depths

Table 2. Example of some gravity gradient interpretation methods.

5. Gravity gradiometry

In the last decade, there has been a new way of measuring gravity developed, not only one component as previous conventional gravity field in the vertical direction $g_{z,}$ but also full

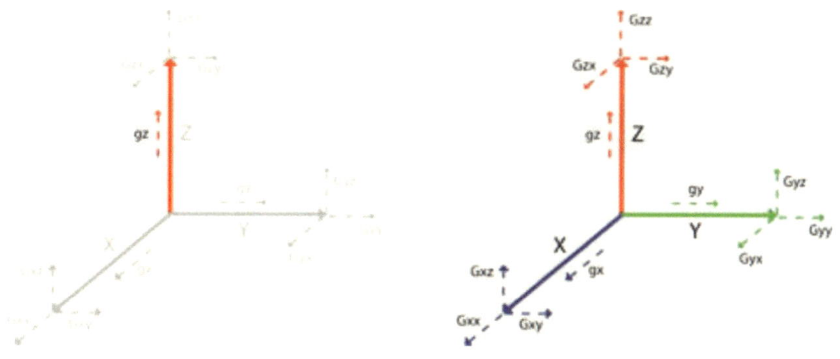

Figure 40. Gravity gradiometry tensors.

tensor gravity gradiometry for all components of the gravity as shown in **Figure 40**. This feature makes the gravity gradient anomaly more localized to the geological source than the gravity anomaly.

6. Gravity data and software

There are many computer programs and codes (**Table 3**) developed by commercial companies or university researchers for gravity data analysis (data filters, data preprocessing, 2D/3D modeling, and 2D/3D inversion).

Software/Code	Link (website)	Capability
PyGMI	http://patrick-cole.github.io/pygmi/index.html#	3D gravity modeling
International Gravimetric Bureau	http://bgi.omp.obs-mip.fr/data-products/Gravity-Databases/Land-Gravity-data	Land and marine gravity data
Fatiando a terra	http://www.fatiando.org/dev/index.html	Modeling and inversion of gravity data
UBC Geophysical Inversion Facility	https://gif.eos.ubc.ca/software/main_programs	Modeling and inversion of gravity data
Geosoft Oasis Montaj	http://www.geosoft.com/products/oasis-montaj	Modeling and inversion of gravity data
Petrel, Gravity Magnetic Modeling and Inversion plug-in (Schlumberger)	https://www.ocean.slb.com/en/plug-ins/plugindetails?ProductId=PGDM-B1	Modeling and inversion of gravity data

Table 3. List of accessible data and computer codes/software for gravity modeling and inversion. There are many other available computer codes for modeling and inversion. The reader can search them at international geophysics journal websites.

7. Conclusions

This chapter summarizes in general from the basics of gravity to instruments used for gravity measurement, gravity corrections, and applications of the gravity method in geosciences. As we stated, gravity is used in many fields of geosciences (geology, structural geology, natural resource exploration, mineral resources, volcanology, geothermal, hydrogeology, CO_2 sequestration, ground cavities, and so).

With the recent development of new gravimeters and availability of satellite gravity data (such as GRACE), the gravity method will be applied in many other new fields for monitoring and estimation of mass changes underground at regional scales (e.g., regional intercontinental aquifers). Also integration and coupling of gravity with other geophysical methods such as magnetics, electromagnetics, seismic and satellite radar such as InSAR will help us increase the accuracy of our geological models and decrease uncertainties in seismic and to help us construct robust geomechanical model for underground oil and gas reservoirs.

Finally, the development of computers and application of robust numerical techniques will certainly help geoscientists to construct near-real geological models.

Acknowledgements

The author acknowledges the financial support of United Arab Emirates University (UAEU) by SURE PLUS grant number: 31S273 (2017). The author acknowledges the following staff and students of UAEU for support in gravity field survey at Al-Ain city: A. Gabr, H. A. Algunaid, A. Q. Al Muntaser, M. Jowhar, F. S. Mohamed. The author thanks Dr. M. Amrouche (Schlumberger, Tokyo Office) for support in 3-D inversion of gravity data. The author thanks Dr. Robert W. Avakian (Oklahoma State University Institute of Technology, USA) for reviewing the chapter. The author would like also to thank Dr. J. Nishijima (Kyushu University, Japan) for absolute gravity data.

A. Appendix

	1st Measurement	2nd Measurement
Station Name		
Station No.		
Altitude	m	
Height	cm	cm
Reading	mGal	mGal
Read Time	sec	
Date	20.. / /	
Time	: :	: :
Earth Tide Correction	mGal	mGal
Data Rejection		
Standard Deviation	mGal	mGal

	1st Measurement	2nd Measurement
Temperature		
Tilt	X arcsec	arcsec
	Y arcsec	arcsec
Operator		
Companion		
Weather		
Wind		
Temperature	°C	°C
Atmospheric Pressure	hPa	hPa

Table A1. Data sheet for gravity measurement using CG-3/CG-3M gravimeters.

B. Appendix

Date:	20.. / ……./……. (……….)
Operator's Name:	
Name of the Station:	
Characteristics of the Battery	
Voltages of the Battery (vehicle battery, AC):	(Volts)
Characteristics of the A-10 Absolute Gravimeter before Measurement:	
"Dropper" Temperature:	(°C) at; (°C) at
"Laser" Temperature:	(°C) at; (°C) at
"IB" Temperature:	(°C) at; (°C) at
Voltages of the Ion Pump:	(Volts) at; (Volts) at
Electric Current of the Ion Pump:	(Ampere)
Voltages of the Fringe amplitude:	(Volts)
Voltages of the Superspring Position SSPOS:	(Volts) at; (Volts) at
*SS POS SS ZERO SS SERVO Please wait 10 min and check again the spring position.	
Characteristics of the station of Measurement	
Longitude:	
Latitude:	
Topographic Elevation:	(m)
Polar motion:	

Date:	**20.. / ……./……. (……….)**
X:	
Y:	
Height:	(cm)
Measured Absolute Gravity:	(μGal)
Weather Conditions (windy, precipitation, cloudy, atmospheric temperature) and Other Remarks (mechanical noise, etc.):	

Table B1. A-10 absolute gravity measurement data sheet.

Author details

Hakim Saibi

Address all correspondence to: saibi.hakim@gmail.com

Department of Geology, College of Science, United Arab Emirates University, Al-Ain, United Arab Emirates

References

[1] Zeng H, Wan T. Clarification of the geophysical definition of a gravity field. Geophysics. 2004;**60**(5):1252-1254

[2] Hunt TM. Gravity changes at Wairakei geothermal field, New Zealand. Geological Society of America Bulletin. 1970;**81**:529-536

[3] Saibi H, Nishijima J, Ehara S. Reservoir monitoring by repeat microgravity measurement at Obama geothermal field, southwestern Japan. Geothermal and Volcanological Research Report of Kyushu University. 2005;**1**(14):27-31

[4] Battaglia M, Gottsmann J, Carbone D, Fermandez J. 4D volcano gravimetry. Geophysics. 2008;**73**(6):WA3-WA18

[5] Saibi H, Gottsmann J, Ehara S. Post-eruptive gravity changes from 1999 to 2004 at Unzen volcano (Japan): A window into shallow aquifer and hydrothermal dynamics. Journal of Volcanology and Geothermal Research. 2010;**191**(1-2):137-147

[6] Reynolds JM. An Introduction to Applied and Environmental Geophysics. John Willey & Sons Edition; New Jersey, USA. 1998. p. 796

[7] Saibi, H. Personal Data

[8] Schmerge D, Francis O. Set standard deviation, repeatability and offset of absolute gravimeter A10-008. Metrologia. 2006;**43**:414-418. DOI: 10.1088/0026-1394/43/5/012

[9] Matsumoto K, Sato T, Takanezawa T, Ooe M. GOTIC2: A program for computation of oceanic tidal loading effect. Journal of the Geodetic Society of Japan. 2001;**47**:243-248

[10] Talwani M, Worzel JL, Landisman M. Rapid gravity computations for two-dimensional bodies with applications to the Mendocino submarine fracture zone. Journal of Geophysical Research. 1959;**64**:49-59

[11] Pirttijärvi M. GRABLOX2. 2014. Available from: https://wiki.oulu.fi/display/~mpi/Gravity+inversion+using+block+model+2. [Accessed: 2017-09-20]

[12] Allis RG, Hunt TM. Analysis of exploitation-induced gravity changes at Wairakei geothermal field. Geothermics. 1986;**51**:1647-1660

[13] Sugihara M, Ishido T. Geothermal reservoir monitoring with a combination of absolute and relative gravimetry. Geophysics. 2008;**73**(6):WA37-WA47

[14] Nishijima J, Saibi H, Sofyan Y, Shimose S,Fujimitsu Y,Ehara S, Fukuda Y, Hasegawa T,Taniguchi M. Reservoir monitoring using hybrid micro-gravity measurements in the Takigami geothermal field, central Kyushu, Japan. In: Proceedings of the World Geothermal Congress; 25-30 April 2010; Bali (Indonesia):2010. p. 1-6

[15] Harnisch G, Harnisch M. Hydrological influences in long gravimetric data series. Journal of Geodynamics. 2006;**41**(1-3):276-287

[16] Crossley D, Xu H, Van Dam T. Comprehensive analysis of 2 years of data from Table Mountain, Colorado. In: Ducame Band Paquet P (eds.) Proceedings of the 13th International Symposium on Earth Tides. Observatoire Royal de Belgique; 1998. p. 659-668.

[17] Gottsmann J, Battaglia M. Deciphering causes of unrest at collapse calderas: Recent advances and future challenges of joint gravimetric and ground deformation studies. In: Gottsmann J, Martí J, editors. Caldera Volcanism: Analysis, Modelling and Response. Elsevier; Amsterdam, Netherlands. 2008. p. 417-446

[18] Saibi H, Nishijima J, Aboud E, Ehara S. Integrated gradient interpretation techniques for 2D and 3D gravity data interpretation. Earth, Planets and Space. 2006;**58**(7):815-821. DOI: 10.1186/BF03351986

[19] Saibi H, Nishijima J, Aboud E, Ehara S. Euler deconvolution of gravity data in geothermal reconnaissance; the Obama geothermal area, Japan. Journal of Exploration Geophysics of Japan. 2006;**59**(3):275-282

[20] Saibi H, Nishijima J, Ehara S. Processing and interpretation of gravity data for the Shimabara peninsula area, South-western Japan. Memoirs of the Faculty of Engineering, Kyushu University. 2006;**66**(2):129-146

[21] Saibi H, Aboud E, Ehara S. Analysis and interpretation of gravity data from the Aluto-Langano geothermal field of Ethiopia. Acta Geophysica. 2012;**60**(2):318-336

[22] Saibi H, Azizi M, Mogren S. Structural investigations of Afghanistan deduced from remote sensing and potential field data. Acta Geophysica. 2016;**64**(4):978-1003 http://link.springer.com/article/10.1515/acgeo-2016-0046

[23] Toushmalani R, Saibi H. Fast 3D inversion of gravity data using Lanczos bidiagonalization method. Arabian Journal of Geosciences. 2015;**8**:4969-4981. DOI: 10.1007/s12517-014-1534-4

[24] Toushmalani R, Saibi H. 3D gravity inversion using Tikhonov regularization. Acta Geophysica. 2015;**63**(4):1044-1065. DOI: 10.1515/acgeo-2015-0029

[25] Farhi W, Boudella A, Saibi H, Bounif MOA. Integration of magnetic, gravity, and well data in imaging subsurface geology in the Ksar Hirane region (Laghouat, Algeria). Journal of African Earth Sciences. 2016;**124**:63-74 http://www.sciencedirect.com/science/article/pii/S1464343X16303077

[26] Bersi M, Saibi H, Chabou MC. Aerogravity and remote sensing observations of an iron deposit in Gara Djebilet, southwestern Algeria. Journal of African Earth Sciences. 2016;**116**:134-150. DOI: 10.1016/j.jafrearsci.2016.01.004 http://www.sciencedirect.com/science/article/pii/S1464343X16300048

[27] Klingele EE, Marson I, Kahle HG. Automatic interpretation of gravity gradiometric data in two dimensions: Vertical gradient. Geophysical Prospecting. 1991;**39**:407-434

[28] Miller HG, Singh V. Potential field tilt – A new concept for location potential field sources. Applied Geophysics. 1994;**32**:213-217

[29] Cordell L, Grauch VJS. Mapping basement magnetization zones from aeromagnetic data in the San Juan Basin, New Mexico, in Hinze, W. J., Ed., the utility of regional gravity and magnetic anomaly maps. Society of Exploration Geophysicists. Oklahoma, USA. 1985:181-197

[30] Reid AB, Allsop JM, Granser H, Millett AJ, Somerton JW. Magnetic interpretation in three dimensions using Euler deconvolution. Geophysics. 1990;**55**(1):80-91

Chapter 4

Application of BEMD in Extraction of Regional and Local Gravity Anomalies Reflecting Geological Structures Associated with Mineral Resources

Yongqing Chen, Binbin Zhao, Jingnning Huang and Lina Zhang

Additional information is available at the end of the chapter

http://dx.doi.org/10.5772/intechopen.71222

Abstract

The bi-dimensional empirical mode decomposition (BEMD) method is an adaptive analysis method for nonlinear and nonstationary data. With the sifting process of BEMD, the data can be decomposed into a series of bi-dimensional intrinsic mode functions (BIMFs), which may present the relative local feature of the data. In this study, the BEMD method was successfully used for analyzing the Bouguer gravity data of Gejiu tin-copper polymetallic ore field in Yunnan Province and Tongshi gold field in Western Shandong Uplift Block to extract different-scale anomalies. In these two cases, regional and local components were separated, which can reflect the geological structures at different depths and some intrusive bodies which may be associated with mineral deposits. The results reveals the spatial distribution relationship between the different intrusive bodies and the various types of mineral deposits in the aforementioned two study area, which provide some reliable evidence for exploration of new concealed mineral deposits.

Keywords: BEMD, gravity anomaly, Gejiu tin-copper polymetallic ore field, Tongshi gold field

1. Introduction

With exhaustion of near surface mineral resources, we are forced to explore deeply for discovering new mineral resources. The exploration and evaluation of concealed mineral deposits represent a frontier in exploration for which geophysical methods are particularly well suited. Geophysical data from gravity and magnetic surveys may contain ore-forming information from greater depths which has not been fully explored due to limitation of traditional mathematical model for extracting useful information. It is possible to extract deeper information from gravity and magnetic data by using new data processing technologies to construct more accurate geological structure models, including the dip and shape of geological bodies, which are essential in predicting deep ore targets. An effective use of gravity fields is to characterize forms, sizes, and depths, as well as masses of various geological objects and their relations to mineral resources. The most direct information acquired from gravity fields is the density of geological bodies. The high gravity indicates higher average densities than the others surrounded. Conversely, the low gravity represents low average densities.

Because of the heterogeneities in the densities of geological bodies during complicated geological processes, multiple solutions are an important problem to be focused. This nonunique correspondence will lead to difficulties in inferring deep geological structures and in delineating concealed geological objects, which need to be constrained by the intrinsic information from geology and geochemistry. On the other hand, some new efficient decomposition are required to extract more accurate anomalies information reflecting geological bodies and obtaining ore-finding evidences from field data.

The recorded data in geosciences usually exhibit nonlinearity and nonstationary due to the complexities of geological processes and the heterogeneities of geological bodies in composition and texture. The traditional methods, such as Fourier transform, cannot analyze nonlinear and nonstationary data seriously. For example, the methods based on Fourier transform are only suitable to analyze linear and stationary data. To handle this problem, Huang developed a new adaptive method for analyzing nonlinear and nonstationary data, named Hilbert-Huang transform (HHT) including empirical mode decomposition (EMD) and Hilbert spectral analysis [1]. The principle of bi-dimensional empirical mode decomposition (BEMD) developed on Huang's basis is elucidated in this chapter. A BEMD program on MATLAB platform was compiled and effectively used to handle regional and local gravity signals for the study areas. This yielded two types of significant two-dimensional intrinsic mode functions (BIMFs) characterizing gravitational components: RBIMFs and LBIMFs. The former depicts spatial distribution relationships between regional geological architectures at different depths which may reflect mantle and/or the basement swells and intrusive bodies which may be associated with mineral deposits. The latter does the spatial distribution relationships between the different intrusive bodies and the various types of mineral deposits, which provide some reliable evidences for searching for new concealed mineral deposits in study areas.

2. Principle of bi-dimensional empirical mode decomposition

The complicated data set can be decomposed adaptively into a finite number of different frequency components called two-dimensional intrinsic mode function (BIMF). These BIMFs are obtained through a sifting process which will not stop until the number of extrema of maximum (or minimum) points is equal or less than 2. Using the BEMD method, the local high frequency oscillations can be extracted from original data set [1].The process of BEMD is similar to the one-dimensional EMD. Obviously, the two-dimensional extrema detection and the surface interpolation of envelops are more complicated than one-dimensional ones.

If *Ori(m,n)* is defined as the original data set which can be decomposed into a finite number of bi-dimensional intrinsic mode functions (BIMFs) from high frequency to low frequency, *Ori(m,n)* can be decomposed into a series of BIMFs and a residue:

$$Ori(m, n) = \sum_{i=1}^{t} B_i(m, n) + Res(m, n) \tag{1}$$

where the $B_i(m,n)$ is the *i*th BIMF component (from high to low frequency, $B_1(m,n)$, $B_2(m,n)$,..., $B_t(m,n)$), and $Res(m, n)$ is the residue. The frequency of $BIMF_1$ is higher than the other BIMFs'.

In the bi-dimensional sifting process, the neighboring window is used for detecting the extrema and the multiquadric method is used for computing the surface interpolation of envelopes to extract the BIMF [2, 3]. For each BIMF, a stopping criterion could be determined in the sifting process. Huang et al. [4] and Nunes et al. [2, 3] proposed that this could be realized by limitation of the size of the standard deviation, *SD*, which can be calculated based on the following formula (2):

$$SD_{ij} = \sum_{m=1}\sum_{n=1} \frac{\left| h_{j(i-1)}(m, n) - h_{ji}(m, n) \right|^2}{h^2_{j(i-1)}(m, n)} \tag{2}$$

The 2D sifting process can be influenced by the determination of *SD*. The number of BIMFs will increase with smaller *SD*. Besides, the mean value and standard deviation of the mean matrix *mean(m,n)* of the upper and lower envelopes should be considered together. The 2D sifting process algorithm used in this study, first proposed by Nunes et al. [2, 3] and improved on the envelope interpolation in this study, is as follows (**Figure 1**):

(1) Initialization: $r_0(m, n) = Ori\ (m, n)$, and $j = 1$ is the BIMF index;

(2) Extraction of the *j*th BIMF component:

   (i) Initialize and make $h_0(m, n) = r_{i-1}(m, n)$, $i = 1$;

   (ii) detect all the points of local maximum ($m_{max,i-1}$) and local minimum ($m_{min,i-1}$) of $h_{i-1}(m, n)$ respectively;

   (iii) compute the upper (lower) envelope of the local maximum and local minimum point, respectively;

   (iv) calculate the envelope mean: $m_{i-1}(m, n) = (m_{max,i-1}(m, n) + m_{min,i-1}(m, n))/2$;

   (v) $h_i(m, n) = h_{i-1}(m, n)–m_{i-1}(x, y)$, $i = i + 1$;

   (vi) $h_i(m, n) = BIMF_j(m, n)$ if $h_i(m, n)$ matches the stopping criterion, then the *j*th BIMF is got, or

   (vii) repeat steps (ii)–(vi);

(3) $r_j(m, n) = r_{j-1}(m, n) - \mathrm{BIMF}_j(m, n)$; and

(4) Go to step (2), when the extrema number of $R_i(m, n)$ is more than 2, $j = j + 1$, or decomposition is ended.

Nunes et al. [2, 3] developed firstly the BEMD algorithm. In this study, the envelope interpolation was improved on the Nunes' method. The two-dimensional sifting process algorithm is illustrated as follows (**Figure 1**).

Huang et al. [1] defined the orthogonality of one-dimensional IMFs using any two IMFs, C_f and C_g through calculating the index of orthogonality (IO) based on the following formula:

$$IO_{fg} = abs\left(\sum_t \frac{C_f C_g}{C_f^2 + C_g^2} \right) \tag{3}$$

For the one-dimensional EMD method, orthogonality is supposed satisfied, if the IO value is less than 0.05.

In this study, we consider two IMFs as vector, and if they are orthogonal, the inner product is zero, that is, $IMF_i(t) \cdot IMF_j(t) = 0$. As is similar to one-dimensional EMD, for BEMD, if the index of orthogonality (IO) value is less than 0.05, we think the orthogonality is satisfied. The IO usually depends on the length of data. The shorter the length is, the bigger the IO is. Here, if we consider an m × n matrix to an m × n vector, the two-dimensional orthogonality is similar to the one-dimensional orthogonality. Here, we say this as a pseudo orthogonality for bi-dimensional data. That is:

$$IO = abs\left(\sum_{p=1}^{m} \sum_{q=1}^{n} \frac{BIMF_i(p, q)\, BIMF_j(p, q)}{BIMF_i^2(p, q) + BIMF_j^2(p, q)} \right), \tag{4}$$

where $BIMF_i(p,q)$ and $BIMF_j(p,q)$ are BIMF components. This method will decrease the error caused by the shortage of data in a given direction.

High-pass (S_{HP}), band-pass (S_{BP}) and low-pass (S_{LP}) defined by Freire and Ulrych [5] can be established in BEMD by selecting one or a few specific BIMF(s) to highlight certain features of geological bodies.

$$S_{HP}(m, n) = \sum_{i=1}^{k} B_i(m, n) \tag{5}$$

$$S_{BP}(m, n) = \sum_{i=k+1}^{p} B_i(m, n) \tag{6}$$

$$S_{LP}(m, n) = \sum_{i=p+1}^{t} B_i(m, n) + Res(m, n) \tag{7}$$

http://dx.doi.org/10.5772/intechopen.71222

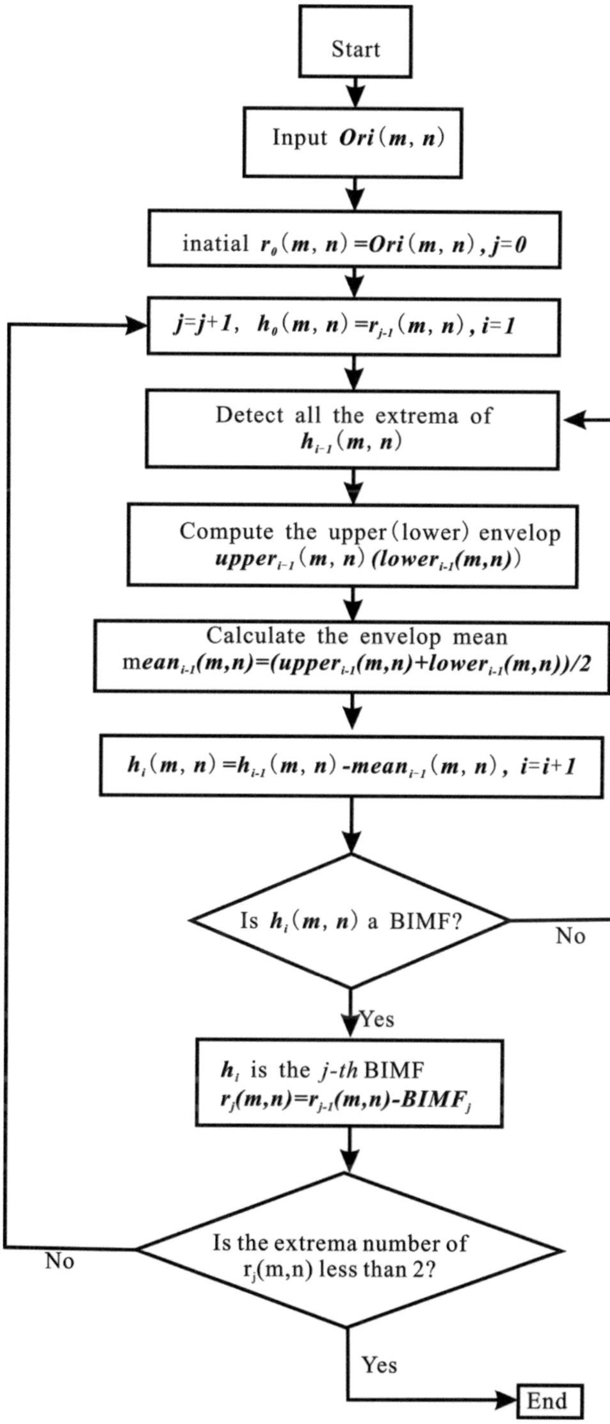

Figure 1. Sifting procedure of BEMD (revised after [2, 3]).

3. Application

3.1. Application of BEMD for extracting gravity anomalies indicating the geological architectures and granites associated mineralization in the Gejiu tin-copper polymetallic ore field

The Gejiu tin-copper polymetallic ore field, located at the westernmost end of the Cathaysia Block in South China, is one of the largest tin-polymetallic ore fields in the world. It is associated with magmatic-hydrothermal ore-forming processes triggered by deeply seated structures and concealed granites (**Figure 2**).

3.1.1. Geological background and mineralization characteristics

The world class Gejiu tin-copper polymetallic ore field is located on the western margin of the South China Block. Bounded on the northwest side by the NE-trending Mile-Shizong Fault, it is adjacent to the Yangtze Block, and bounded on the southwest side by the NW-trending ASRR Fault, it is adjacent to the Indo-China Block (**Figure 2**). The Gejiu ore field is approximately 1600 km^2 in area and was a tectonic sag or depression for much of its geological

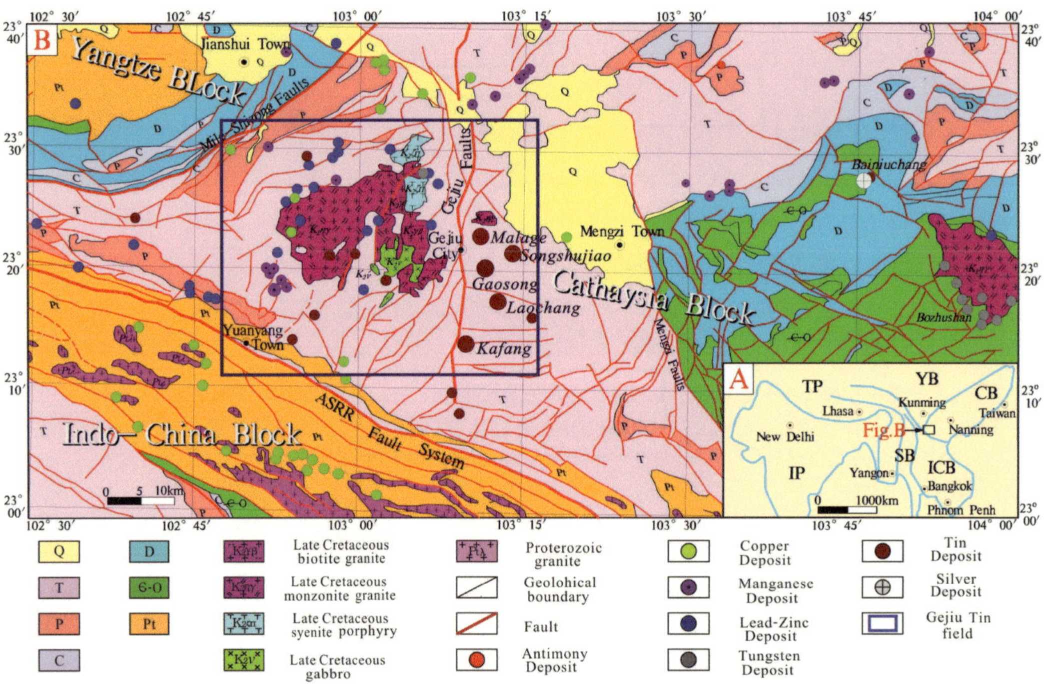

Figure 2. A: Geological outline map of southeastern Asia, showing major tectonic units and location of the Gejiu tin-copper polymetallic ore field is situated (from [6]); and B: simplified geological map showing the spatial distribution of intrusive rocks and the various types of mineral deposits in the Gejiu tin-copper polymetallic ore field and surrounding areas. CB, Cathaysia Block; YB, Yangtze Block; SB, Sibumasu Block; TP, Tibet Plate; ICB, Indo-China Block; and IP, India Plate.

history. Cambrian to Quaternary rock successions are well preserved, and the late Triassic to Cretaceous strata are preferentially exposed at the surface due to uplift associated with Yanshanian (Mesozoic) tectonic movements [7]. Most of the outcrops in the Gejiu area consist of Middle Triassic Gejiu Formation carbonates with more than 3000 m in thickness, and Middle Triassic Falang Formation fine-grained clastic sediments and carbonates with interlayering mafic lavas ranging from 1800 to 2800 m in thickness [8]. Numerous faults exist in the Gejiu area, including the NNE-trending and NE-trending faults and the NS-trending Gejiu fault (**Figure 2**). The Gejiu fault divides the Gejiu ore field into two parts (**Figure 2b**). The ore deposits are mainly located in the eastern part. Structures in the eastern part include NS-, NNE- and EW-trending faults. The EW-trending faults separate the Gejiu field into five segments from south to north, that is, Kafang, Laochang, Gaosong, Songshujiao, and Malage (**Figure 2b**). Besides Triassic lavas, the Mesozoic igneous rocks in this area can be found, for example gabbro, mafic microgranular enclave (MME)-bearing granites, porphyritic biotite granite, equigranular biotite granite, alkaline rocks and mafic dikes. SHRIMP and LA-ICP-MS zircon ages ranging from 77.4 ± 2.5 to 85.8 ± 0.6 Ma have been reported for the granitic rocks [9, 10]. The granites are transitional from metaluminous to weakly peraluminous, and are high-K and alkali-rich. Geochemical and Sr–Nd–Hf isotopic data suggest that the Gejiu granitic magma experienced high degrees of fractional crystallization following its formation due to partial melting of the Mesoproterozoic crust with minor input from mantle-derived magmas. The geochemical data indicate that the degrees of fractional crystallization are distinctive for different intrusive phases and that the later phases tend to be more evolved. These highly fractionated granites were likely generated in an intraplate setting. Mineralization is spatially related to highly evolved granitic phases. The previous studies generally suggest that the mineralization was coeval with Late Cretaceous granitic intrusions [9, 10]. The deposits in the Gejiu ore field are mostly underlain by the mid-Triassic Gejiu Formation, which consists primarily of light gray to gray limestone, with minor banded and lenticular calcareous dolomite. Greisen (quartz–muscovite), muscovite, sericite and skarn alterations were intensively developed on the margins of the cupola and in the contact zone between granite and carbonate. Skarn-sulfide ores occur along the contact zones between granite and limestone and/or dolomite, and are characterized by disseminated, veinlet and massive structures. The primary skarn minerals are pyroxene, garnet and scapolite. Tourmaline and fluorite are common hydrothermal alteration minerals overprinting skarn ores. The ores are associated with retrograde skarn minerals (actinolite, tremolite and chlorite). The major ore minerals include arsenopyrite, pyrrhotite, cassiterite, chalcopyrite and marmatite, with lesser amounts of pyrite, scheelite, native bismuth, molybdenite and magnetite [6, 11].

3.1.2. Gravity decomposition

The gravity data surveyed at a grid of 2000 m × 2000 m, covering an area including the Gejiu tin-copper polymetallic ore field and the Bozhushan silver-lead-zinc polymetallic ore field (**Figure 3**), are provided by the Yunnan Geological Survey. The total data resolution is $\pm 2.32 \times 10^{-6}$ m/s^2. The densities of the exposed geological bodies vary from 2.55 to 2.60 g/cm^3 for the granites, 2.68 to 2.72 g/cm^3 for the carbonate rocks, 2.98 to 3.10 g/cm^3 for the basic and ultrabasic rocks, and average 4.52 g/cm^3 for massive skarn tin ore within the Gejiu district [12].

In this study, a $SD = 0.02$ is assumed for decomposition of the gravity data surveyed for a grid of 2000 × 2000 m, according to the Eq. (8) using the sifting process of the BEMD described above:

$$Ori(m, n) = \sum_{i=1}^{3} B_i(m, n) + Res(m, n) \quad (8)$$

where $Ori(m, n)$ is the original 2D gravity data; $B_i(m, n)$ is the 2D IMFs; and $Res(m, n)$ is the 2D residual component. The original gravity image (**Figure 3**) was decomposed into three BIMFs ($BIMF_1$, $BIMF_2$ and $BIMF_3$) and one residue $Res(m, n)$ by BEMD. Because $BIMF_1$, $BIMF_2$ and $BIMF_3$ decrease in frequency, within the same range, $BIMF_1$ represents the gravity image with the highest frequency, and $Res(m, n)$ represents the gravity image with lowest frequency. The frequency of both the $BIMF_2$ and the $BIMF_3$ images ranges between that of $BIMF_1$ and $Res(m, n)$. The 1D decomposition results for the northeastern (line AB) and northwestern directions (line CD) of the Gejiu tin-copper polymetallic field at a mutually perpendicular angle (**Figure 3**) are shown in **Figure 4**. The smoothness of the curves increases as the frequency decreases from $BIMF_1$ to residual component. The orthogonality of the BIMFs was checked in both directions (**Table 1**). The results show that except for IO_{23} being slightly bigger than zero, IO_{12} and IO_{13} values are close to zero and thus orthogonality is approximately satisfied.

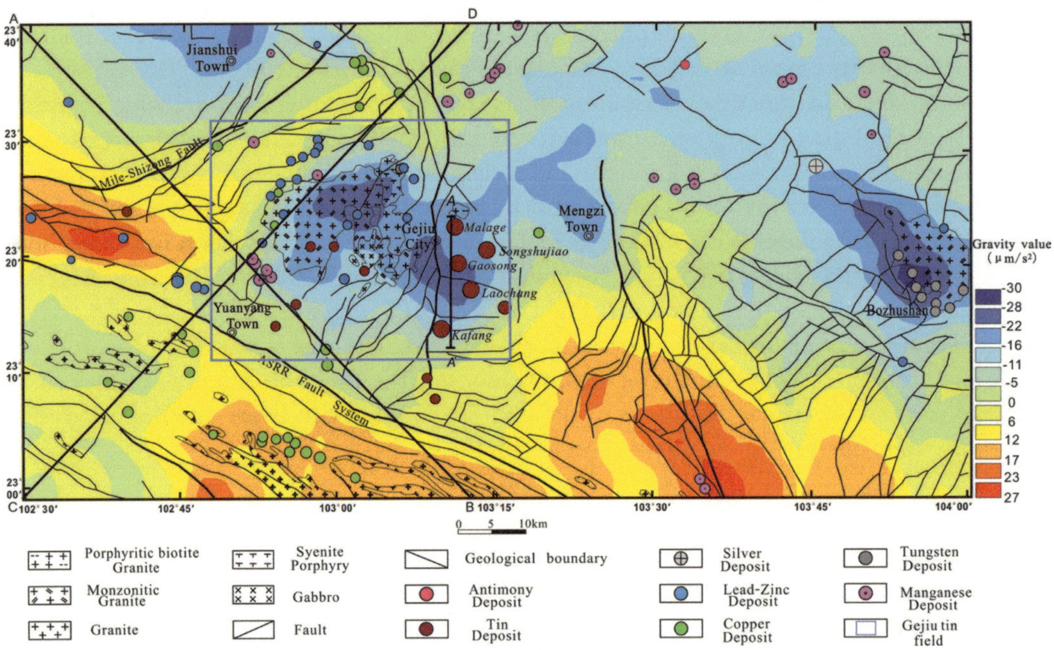

Figure 3. Original gravity image surveyed at a scale of 1:200,000 for the Gejiu tin-copper polymetallic field and surrounding areas.

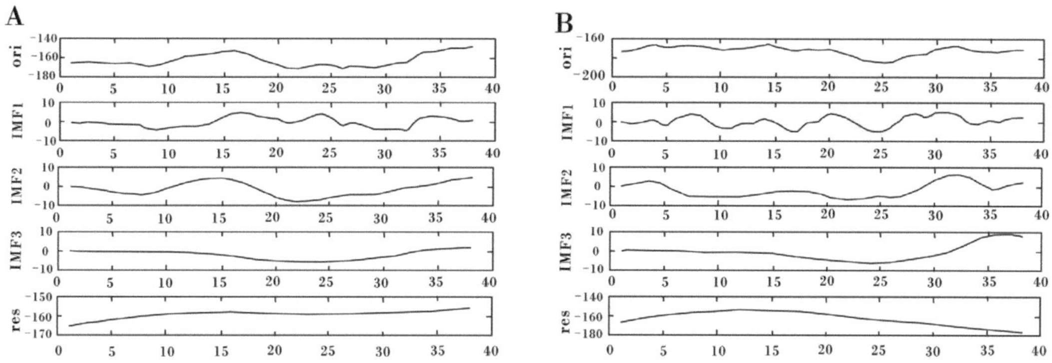

Figure 4. IMF components of 1D gravity data for the NW-trending section AB (A) and NE-trending section CD (B) shown in **Figure 3**.

Section (line)	AB	CD
IO_{12}	0.0466	0.1112
IO_{13}	0.0885	0.058
IO_{23}	0.3607	0.2996

Table 1. Orthogonality assessment.

3.1.3. Implications of the gravity decomposition components for the geology and mineralization in the Gejiu tin-copper polymetallic field

The abovementioned BIMFs have distinct geological implications. The high-pass filtered image ($BIMF_1$) depicts the shallow geological architecture in the study area (**Figure 5**). Three negative gravity anomalies bracketed by positive gravity anomalies are shown (**Figure 5**). Two among the three negative gravity anomalies correspond respectively to the outcropped granites (I) in the western Gejiu ore field and the buried granites (II) in the eastern Gejiu ore field. The positive gravity anomalies are conjectured to be created from faults filled by basic dykes and/or skarn alteration with tin-copper mineralization (IV). At the northwest side of the Gejiu ore field bounded by the NE trending Mile-Shizong fault, there is an obvious negative gravity anomaly (III) that we conjecture could be created by buried granites, which could be a new area for exploring tin-lead-zinc polymetallic deposits. A geological-geophysical profile across the whole Gejiu tin-polymetallic field from north to south (line AA′ in **Figure 5**) was drawn to show the variations of both the $BIMF_1$ gravity component and the original gravity anomaly along the geological profile (**Figure 6**). It is illustrated that the $BIMF_1$ gravity component more exactly reflects the shape fluctuation of the concealed granites, the contact metasomatic skarn alteration and tin-copper polymetallic mineralization than the original gravity anomaly and the local gravity component from

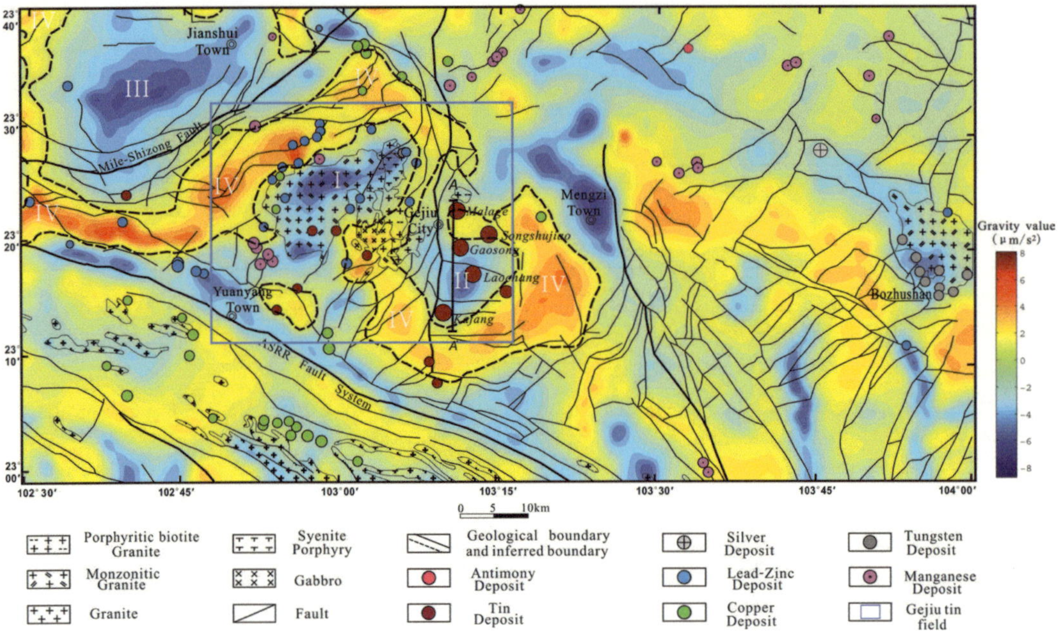

Figure 5. The gravity component BIMF$_1$ image showing that there are three negative gravity anomalies which may be associated with the tin-copper polymetallic mineralization, surrounded by rings of positive gravity anomalies which are inferred to originate from faults filled with skarn alteration and tin-copper polymetallic mineralization.

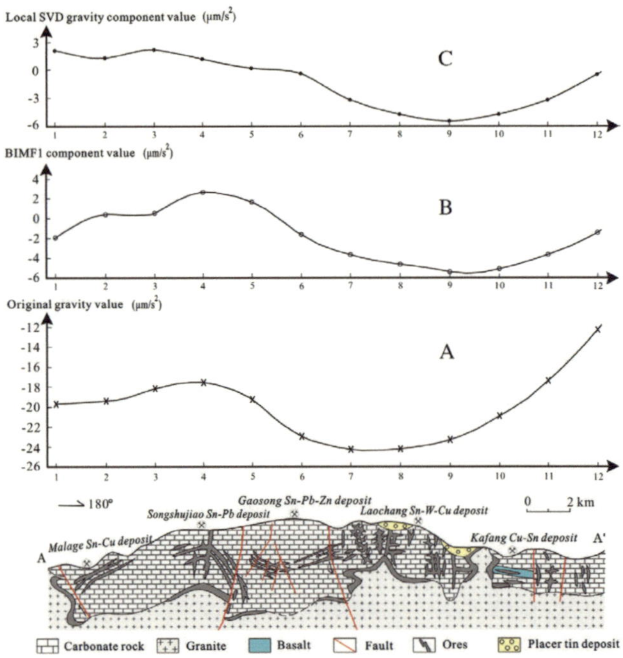

Figure 6. Spatial relationships between granites, ore bodies and local gravity components cross the section AA′ in the Gejiu tin-copper polymetallic ore field shown in **Figure 5** (modified from [10, 11]). (A) Original gravity anomaly; (B) the BIMF$_1$ gravity component decomposed from the original gravity anomaly by using BEMD; (C) local gravity component decomposed from the original gravity anomaly by using SVD [13].

the singular-value decomposition (SVD) [13]. The $BIMF_1$ gravity component values vary from about −3.0 to +2.5 μm/s² from the Malaga Sn-Cu ore block toward the Songshujiao Sn-Pb ore block, in accordance with an increase in thickness of the overlying carbonate rocks, whereas from the Songshujiao Sn-Pb ore block to the Kafang Cu-Sn ore block, the IMF_1 gravity component values range from +2.5 to −5.5 μm/s², exhibiting a sharp decrease corresponding to thinning of the overlying carbonate rocks. At the south end of the section, the $BIMF_1$ gravity component values increase from −5.5 to −2.0 μm/s² in accordance with thickening of the overlying carbonate rocks (**Figure 6**). These observations imply that the local gravity anomaly variation may be mainly influenced by the shape fluctuation of the granite complex and the thickness of the overlying carbonate rock as well as local mineralization. The band-pass filtered image ($BIMF_2$) (**Figure 7**) depicts the middle-shallow geological architecture. There is a negative gravity anomaly centered area which may reflect the overall distribution of the Gejiu granitic complex, including the outcropped and the buried granites. The outcropped granites in the western Gejiu field bounded by the Gejiu fault may extend to the eastern Gejiu field at subsurface, which has been tested by drilling and mining engineering [10]. Almost all types of ore deposits are distributed around the Gejiu granitic complex displaying negative gravity anomaly, and have a close temporal-spatial relationship with the granitic complex [11].The other band-pass filtered image ($BIMF_3$) (**Figure 8**) depicts the middle-lower geological architecture. There is an EW-trending negative gravity anomaly zone (I) with two negative gravity anomaly centers (I_a and I_b) that coincide with the Gejiu tin-copper polymetallic ore field and the Bozhushan silver-lead-zinc polymetallic ore field, respectively [14, 15]. This may reflect the existence of an EW-trending granite zone at middle-low depths within the study area. The Gejiu tin-copper polymetallic ore field is

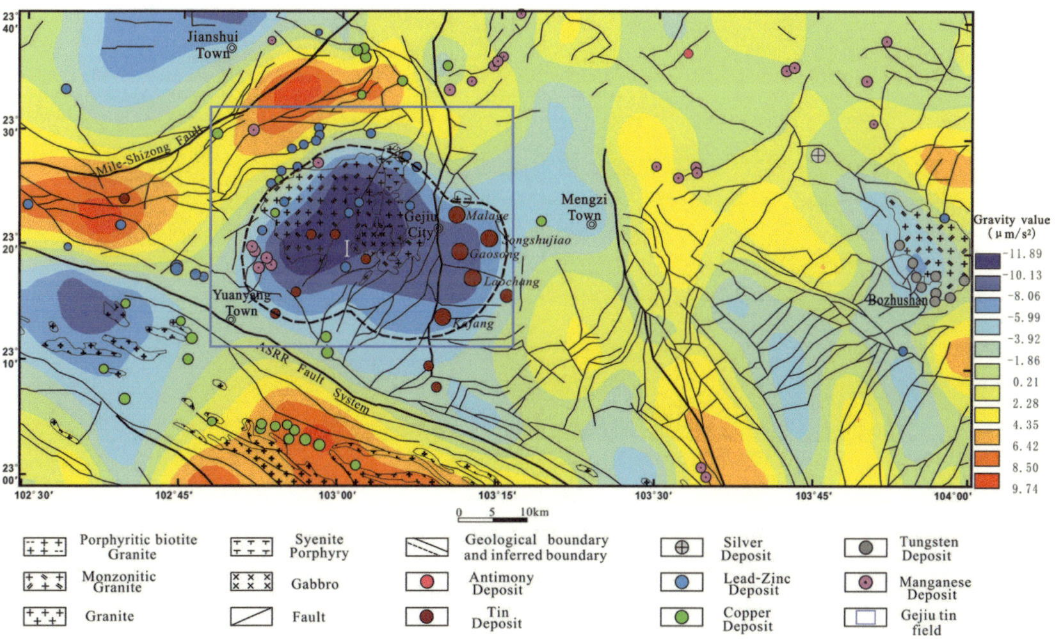

Figure 7. The gravity component $BIMF_2$ image showing a negative gravity centered area which may reflect existence of a huge granitic complex within the Gejiu tin-copper polymetallic ore field. The various ore deposits within the Gejiu tin-copper polymetallic ore field have a close spatial relationship with the granitic complex.

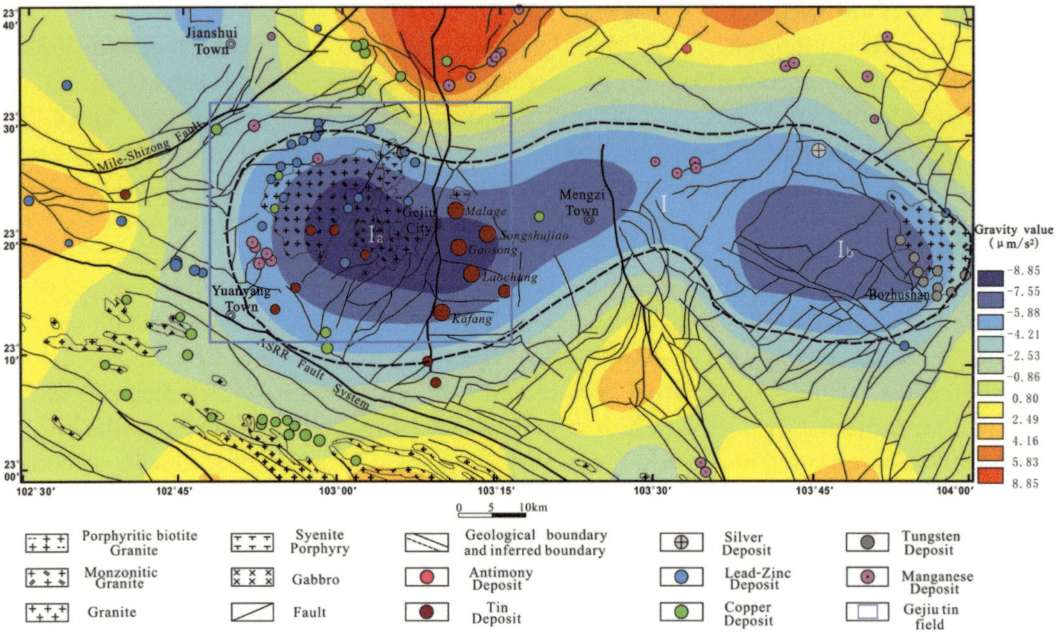

Figure 8. The gravity component $BIMF_3$ image showing an EW-trending negative gravitation anomaly zone that may reflect existence of an EW-trending granite zone connecting the Gejiu tin-copper polymetallic field to the Bozhushan silver-lead-zinc polymetallic field.

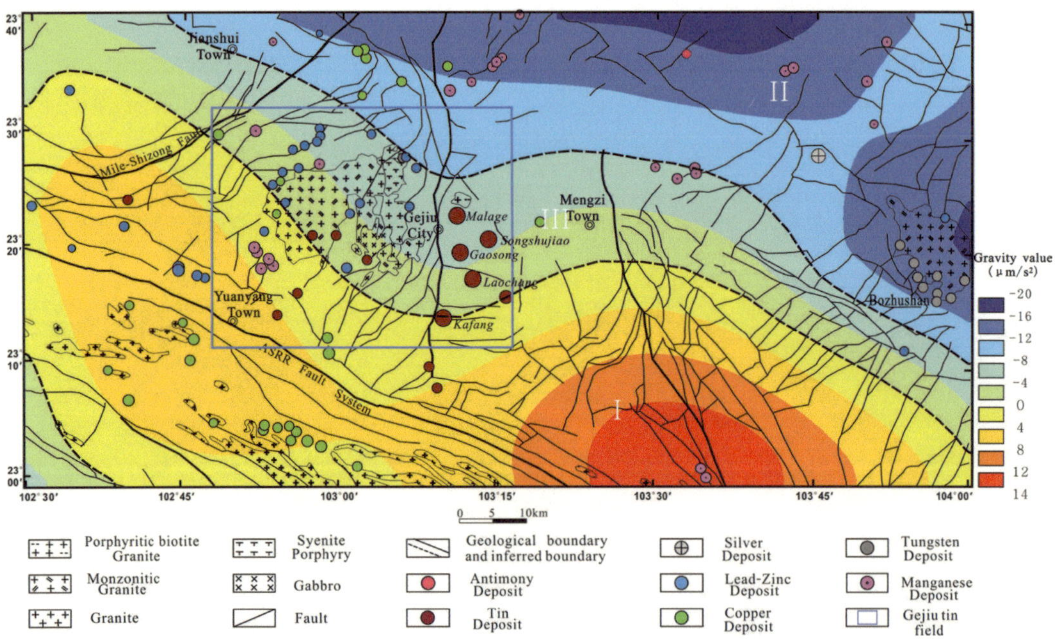

Figure 9. The residual gravity component *Res*(*m*, *n*) image showing the lowest geological architecture within the study area.

connected to the Bozhushan silver-lead-zinc-tin polymetallic ore field by this granite zone, forming an EW-trending regional polymetallic mineralization zone associated with the granite zone (**Figure 8**). The low-pass filtered gravity component image ($Res(m, n)$) (**Figure 9**) may depict the deep geological architecture within the study area, as suggested by Hou et al. [16]). In our study area, the residual map (**Figure 9**) depicts a pair of NWW-trending positive and negative gravity anomaly zones (I and II). The central axis of the positive gravity anomaly zone (I) approximately coincides with the ASRR fault which is regarded a boundary between the Indochina Block and the Cathaysia Block [17–21]. It has been illustrated that the ASRR fault cuts through the crust and controls the regional distribution of both intrusions and related mineralization [22, 23], which may have caused the mantle and/or the basement uplift with distinct positive gravity anomaly [13, 24, 25]. The negative gravity zone is inferred to be associated with the mantle and/or the basement depression (II) [13, 24–26]. The Gejiu tin-copper polymetallic deposits and the related granites that are characterized by initial $^{87}Sr/^{86}Sr$ ratios ranging from 0.7097 to 0.7130 suggesting mixed crust–mantle sources [6], are located at the transitional zone (III), whereas the Bozhushan silver-lead-zinc polymetallic deposits and the related granites with initial $^{87}Sr/^{86}Sr$ ratios ranging from 0.7126 to 0.7170 (unpublished data from authors), are situated within the zone of mantle depression. This difference in magmatic sources may be a key reason why huge amounts of tin and copper were accumulated within the Gejiu tin-copper polymetallic ore field, whereas silver, lead and zinc deposited were concentrated within the Bozhushan ore field.

3.2. Application of BEMD for extracting gravity anomalies indicating the geology and mineralization of the Tongshi gold field, Western Shandong Uplifted Block

3.2.1. Geology setting

The Tongshi gold field is located in the concealed basement to the southwestern the Mesozoic Pingyi volcanic sedimentary basin in the western Shandong Uplifted Block, Eastern China. The concealed basement area, where the Archean greenstone covered by Cambrian–Ordovician carbonates, was intruded by the Tongshi subvolcanic alkaline intrusive complex consisting of syenite porphyries and diorite porphyrites. In the Tongshi gold field, the Tongshi complex is the primary ore-controlling factor. Gold deposits (occurrences) distributed around or in the Tongshi complex can be classified into four types: (a) porphyry gold occurrences are located in the Tongshi complex, (b) Skarn iron–copper–gold deposits are located in the inner contact metasomatic zone between the intrusive complex and its host Cambrian–Ordovician carbonate rocks surrounded, (c) Crypto-breccia gold deposits and (d) Carlin-type gold deposits located in the outer contact metasomatic zone. Thus, the various types of gold deposits constitute a gold mineral resource series (**Figure 10**) [27, 28]. Two $w(^{40}Ar)/w(^{39}Ar)$ analysis of amphiboles from the diorite porphyrite and the syenite porphyry suites revealed ages of 189 and 188 Ma, respectively [29]. The intrusive complex's zircon SHRIMPU–Pb ages vary from 167.9 to 183 Ma [30]. The intrusive complex exhibits obvious negative gravity fields (**Figure 10**).

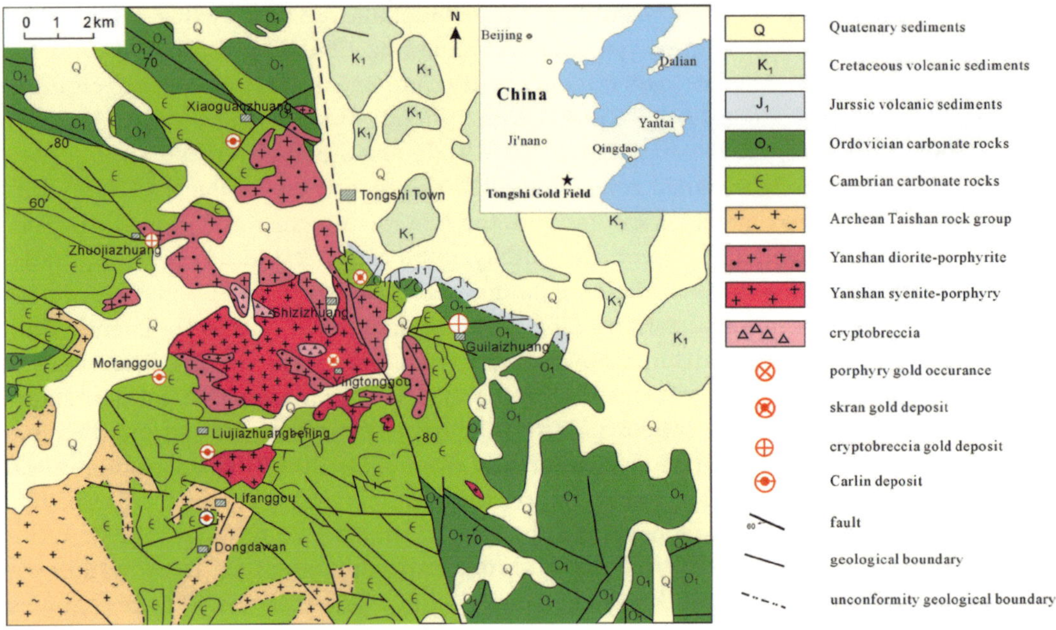

Figure 10. Geology and mineralization of Tongshi gold field, Western Shandong Uplifted Block, Eastern China (revised after [27]).

3.2.2. Gravity components decomposition

Gravity data analyzed here were surveyed at a grid of 500 m × 250 m provided by the Yanzhou Geology & Mineral Exploration Institute of Shandong province, and the study area covered about 408 km^2. The resolution of the survey data is ±2.32 × 10^{-6} m/s^2. The density of exposed geological bodies was 2.73–2.90 g/cm^3 for the Archean greenstone belt, 2.64–2.76 g/cm^3 for Cambrian–Ordovician carbonates, 2.61–2.71 g/cm^3 for the Tongshi intrusive complex and 2.46–2.53 g/cm^3 for volcanic sedimentary rocks [31]. The Bouguer gravity data (**Figure 11**) can be separated to five BIMFs ($BIMF_1$, $BIMF_2$, $BIMF_3$, $BIMF_4$ and $BIMF_5$) and a residual component *Res*(*m*, *n*) with sifting processes

$$Ori(m, n) = \sum_{i=1}^{5} BIMF_i(m, n) + Res(m, n) \tag{9}$$

where Ori(*m*, n) is the original 2D Bouguer anomaly data; $BIMF_i(m, n)$ is the 2D IMFs and *Res*(*m*, *n*) is the 2D residual component. The BIMFs represent 2D gravity data characterizing the local feature with different scale which may reveal geological feature. Though $BIMF_1$, $BIMF_2$, $BIMF_3$, $BIMF_4$ and $BIMF_5$ decrease in frequency, the frequency of $BIMF_i$ is higher than that of BIMF i + 1 only within the same range, not in any range.

Here we set SD = 0.02 for the stopping condition. The 1D decomposition results for the northeastern (line AB) and northwestern directions (line CD) (**Figure 11**) are shown in **Figures 12** and **13**, respectively. The smoothness of the curves increases with the frequency decrease

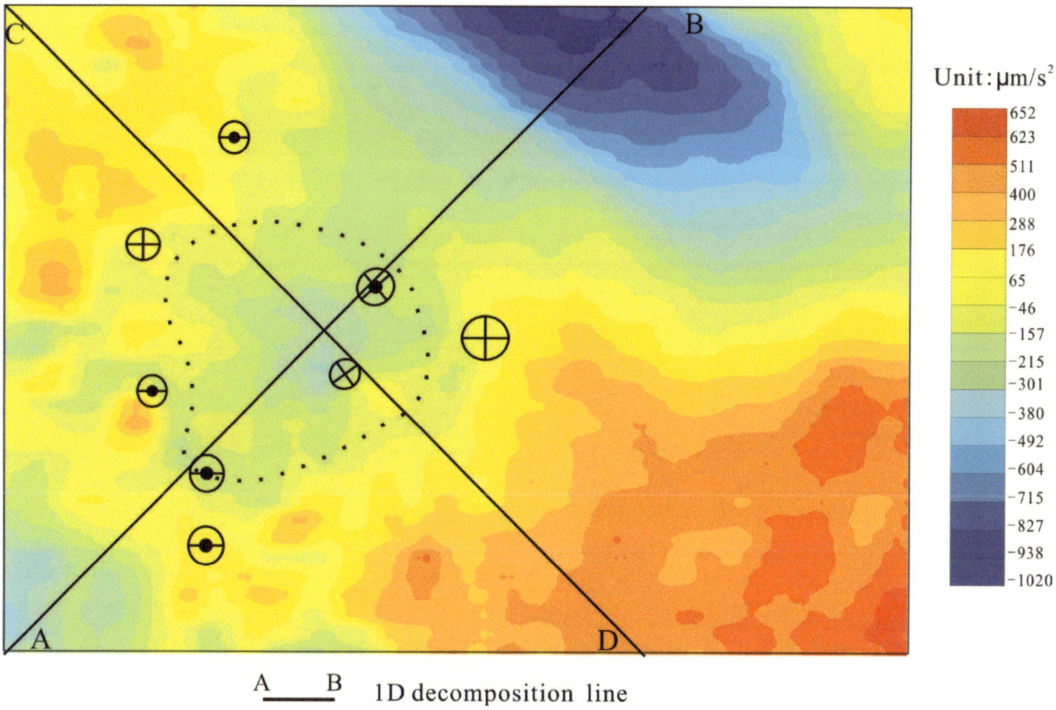

Figure 11. Original Bouguer gravity anomaly image surveyed at a scale of 1:50,000 in Tongshi gold field.

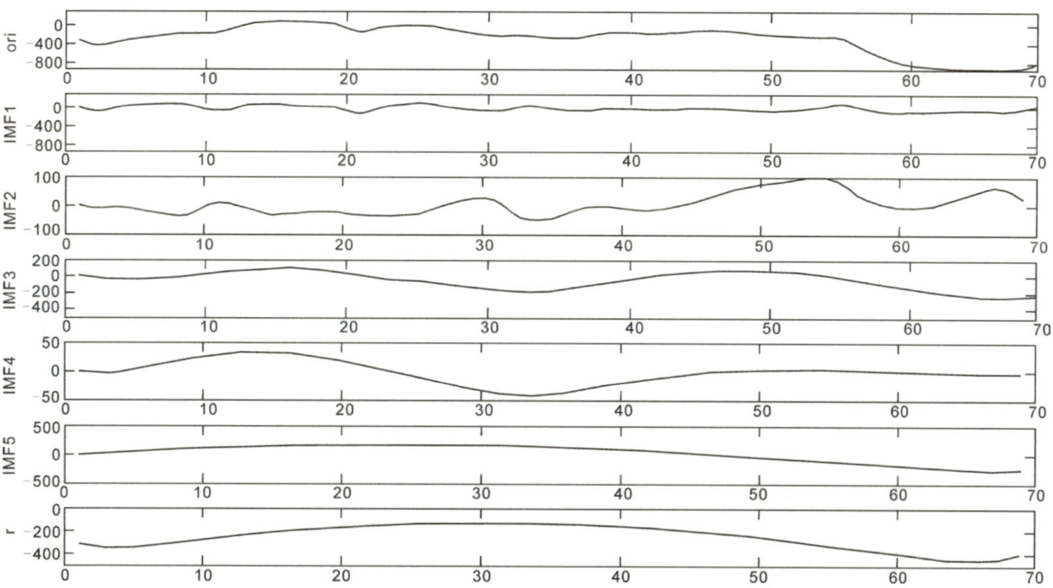

Figure 12. IMF components of one-dimensional gravity data for NE trend of a section AB shown in **Figure 11**.

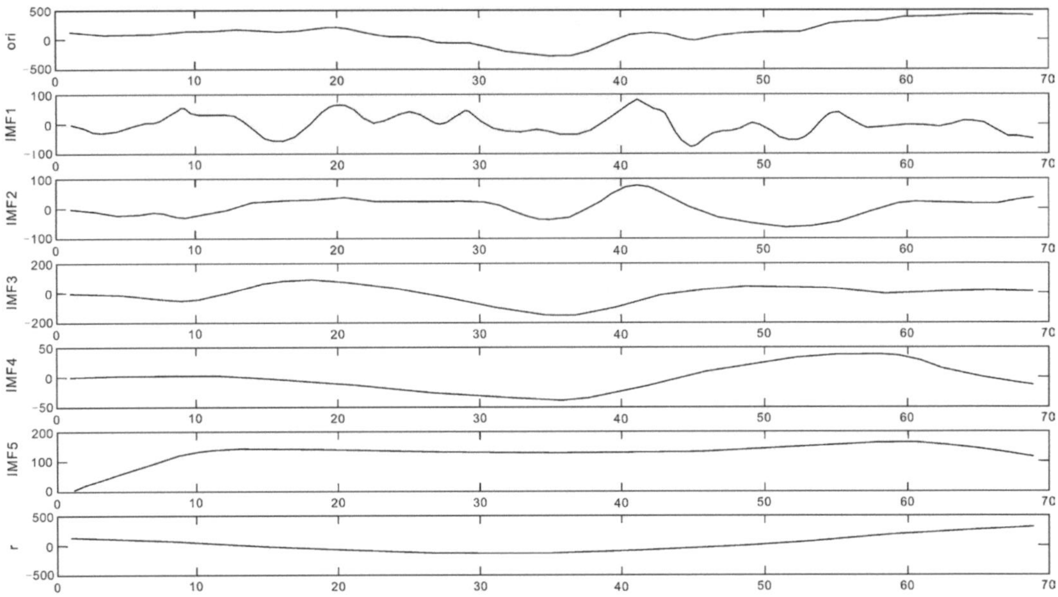

Figure 13. IMF components of one-dimensional gravity data for NW trend of a section CD shown in **Figure 11**.

from IMF_1 to IMF_5. Almost all the *IOs* we calculated are close to zero which approximately satisfied the orthogonality (**Table 2**).

3.2.3. Implications of the gravity decomposition components for the geology and mineralization in the Tongshi gold field

Both $BIMF_5$ and $BIMF_3$ components among the five BIMFs mentioned above have clear geological significance (**Figures 14** and **15**). The $BIMF_5$ image in **Figure 14** is considered as a

	Section AB	Section CD
IO_{12}	0.1008	0.1726
IO_{13}	0.1065	5.6550×10^{-4}
IO_{14}	0.08	0.0506
IO_{15}	0.106	0.0952
IO_{23}	0.0209	0.029
IO_{24}	0.0028	0.0245
IO_{25}	0.1426	0.014
IO_{34}	0.0954	0.0828
IO_{35}	0.1698	0.0664
IO_{45}	0.0111	0.0157

Table 2. Orthogonality assessment.

http://dx.doi.org/10.5772/intechopen.71222

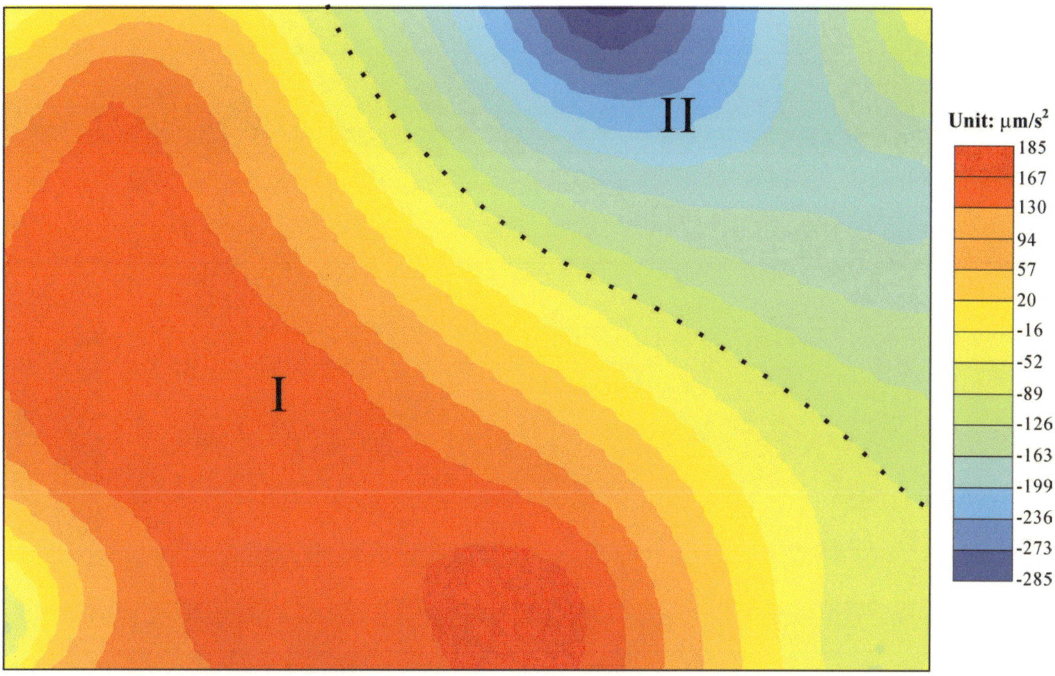

Figure 14. $BIMF_5$ image decomposed from original gravity data for Tongshi gold field.

low-pass-filtering result of the Bouguer anomaly data. Two basic deep geological units are obvious. Unit I, showing positive Bouguer anomaly, is the concealed basement with a double layer comprising Archean greenstone covered by Cambrian–Ordovician carbonate rocks. Unit II, with an obvious negative Bouguer anomaly, is the Mesozoic volcanic sedimentary basin that consists of Jurassic and the Cretaceous volcanic-sedimentary rocks. The $BIMF_3$ image (**Figure 15**) is considered as a band-pass-filtering result, which clearly reveals the geological structure of the Tongshi gold field. The Tongshi gold field includes three basic geological units according to **Figure 10**. The NW Mesozoic volcanic-sedimentary basin (II) developed in the northeast area of the study area present an obvious negative Bouguer anomaly. The Tongshi intrusive complex (III) intruded into the concealed basement exhibits relative negative gravity anomaly comparing to the host rocks.

According to its gravity anomaly features, Unit III can be subdivided into subunits IIIa, obvious negative anomalous center reflecting the distribution of Tongshi intrusive complex, and IIIb, circular positive gravity anomaly representing the contact metasomatic zone between the Tongshi intrusive complex (IIIa) and its host rocks. The spatial distributions of various types of gold deposits are controlled by Unit III (**Figure 15**).

Taking a geological section along the line EF in **Figure 15** yields a geological-geophysical pattern for the Tongshi gold field showing the formation and distribution of gold deposits (**Figure 16**). During 168–189 Ma, the Tongshi intrusive complex (IIIa), with the negative Bouguer anomaly varying from −2 to −265 $\mu m/s^2$, intruded into the concealed basement.

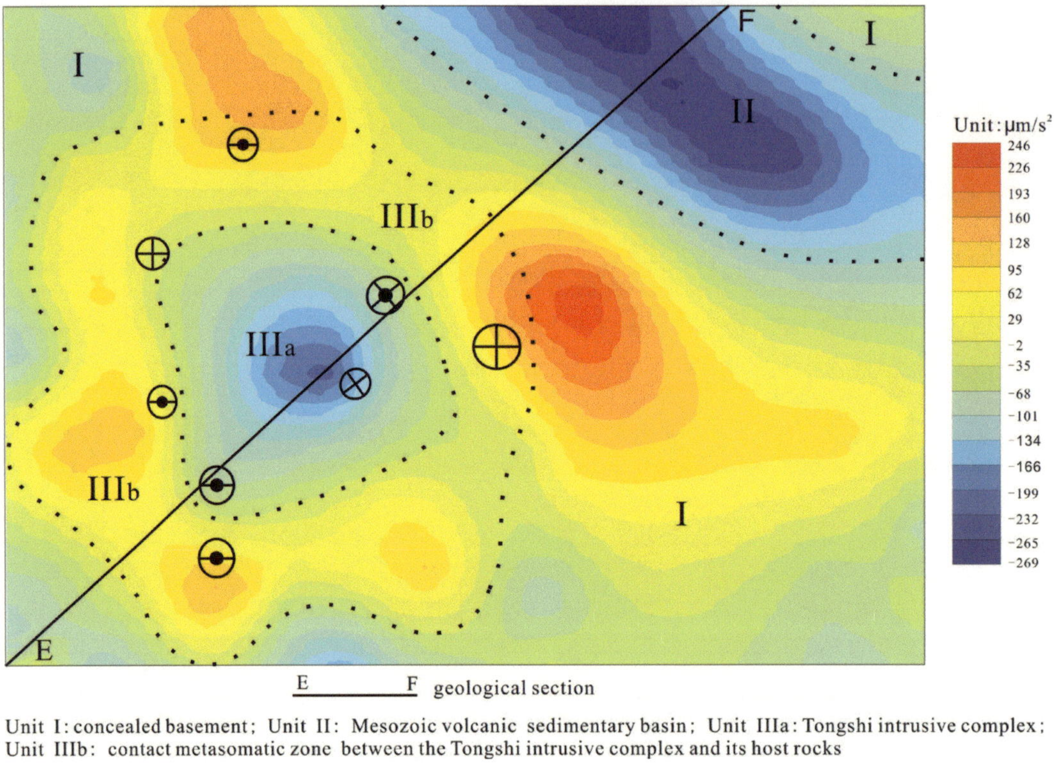

Figure 15. BIMF$_3$ image decomposed from original gravity data for Tongshi gold field.

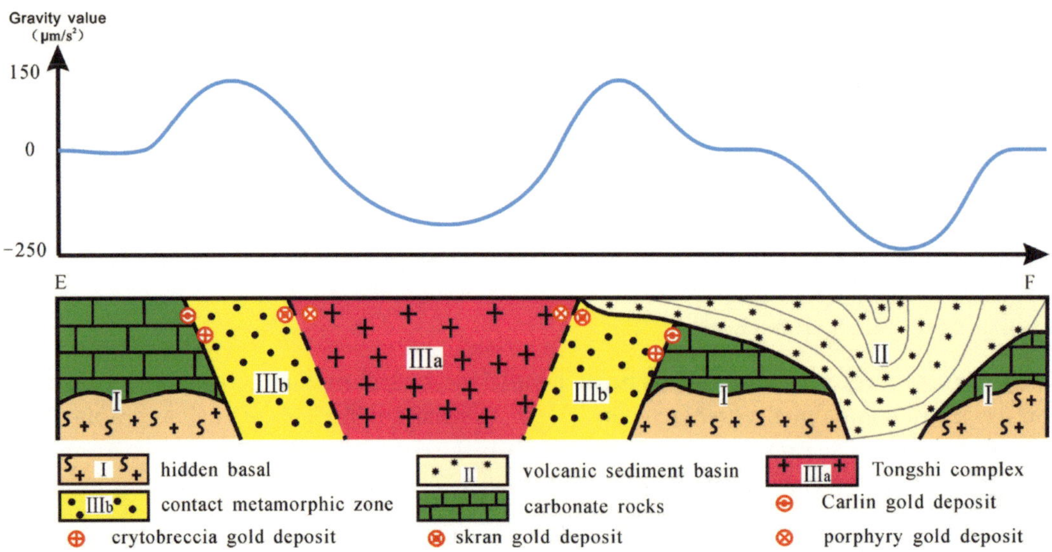

Figure 16. Geological-geophysical model of Tongshi gold field cross section EF shown in **Figure 15**.

The extracted Bouguer anomaly of the concealed basement (I) generally varies from −100 to −2 μm/s^2. But in the northeast margin, it rapidly increases from −2 to 246 μm/s^2, possibly reflecting upwelling zones of the Archean basement. The contact metasomatic zone shows positive gravity anomalies, varying from 0 to 160 μm/s^2 (IIIb), which can be attributed to metasomatic processes when magma intruded upwards into Cambrian–Ordovician carbonates. The distribution of various types of gold deposits is controlled by Unit III and displays spatial zonation. In the Tongshi intrusive complex (IIIa), porphyry gold occurrences can be found, and Skarn iron–copper–gold occurrences are located in the inner contact metasomatic zone (IIIb). Crypto-breccia and Carlin-type gold deposits are located at the outer contact metasomatic zone (IIIb) (**Figure 16**). Some undiscovered concealed gold deposits might be found on the northeastern side of the Tongshi intrusive complex (upwelling zones of the Archean basement) and the contact metasomatic zone covered by early Cretaceous volcanic-sedimentary rocks.

4. Conclusions

BEMD can effectively extract gravity components at different wavelengths that may reflect geological architectures at different depths associated with mineralization and the related intrusions in the study areas. The components are extracted by decomposing the gravity data into different bi-dimensional intrinsic mode functions (BIMFs). The gravity components of $BIMF_1$, $BIMF_2$, and …, $BIMF_i$ and $BIMF_{i+1}$, … decrease in frequency. The residual gravity component (*Res*(*m*, *n*)) has the lowest frequency. These gravity components decomposed by BEMD have the distinct geological implications. For the Gejiu tin-copper polymetallic field: (a) The high-pass filtered gravity component image ($BIMF_1$) depict the shallow geological architecture within the Gejiu tin-copper polymetallic field, which indicates that the skarn alteration and tin-copper mineralization with positive gravity anomalies are distributed around the granites with negative gravity anomalies. (b) The band-pass filtered gravity component images ($BIMF_2$) depict the middle-shallow geological architecture, which indicates that the outcropped granites in the western Gejiu ore field bounded by the Gejiu fault may extend to the eastern Gejiu ore field to form an integrated granitic complex in the subsurface. (c) The other band-pass filtered gravity component images ($BIMF_3$) depict the middle-lower geological architecture, which indicates that there may be an EW-trending granite zone with negative gravity anomaly at middle-lower depth connecting the Gejiu tin-copper polymetallic field to the Bozhushan silver-lead-zinc polymetallic field, displaying the existence of an EW-trending regional polymetallic mineralization zone within the study area. (d) The low-pass filtered gravity component image (the *Res*(*m*, *n*)) depict the lowest geological architecture within the study area, which reflects that there may be a pair of the NW-trending uplift zone of the mantle and/or the basement with positive gravity anomaly and the depression zone of the mantle and/or the basement with negative gravity anomaly. The Gejiu tin-copper polymetallic deposits and the related granites are located at their transitional zone, but the Bozhushan silver-lead-zinc polymetallic deposits and the related granites are situated within the depression zone, which imply that the diversity of the Late Yanshanian granites and the related polymetallic deposits in the study area may be controlled by the complexity of the crust-mantle interaction in depth.

For the Tongshi gold field, both $BIMF_5$ and $BIMF_3$ components among the five BIMFs have clear geological significance (**Figures 14** and **15**). The $BIMF_5$ image in **Figure 14** is a low-pass filtering component of the original Bouguer gravity anomaly data. Two basic deep geological units are obvious. Unit I is a concealed basement with a double layer comprising Archean greenstone with the Cambrian–Ordovician carbonate rocks laid on, characterized mainly by positive gravity anomaly, and Unit II is the Mesozoic volcanic sedimentary basin that consists of Jurassic and the Cretaceous volcanic-sedimentary rocks, with an obvious negative Bouguer gravity anomaly.

The $BIMF_3$ image in **Figure 15** is a band-pass filtering component of the Bouguer gravity anomaly, which reveals the geological structure clearly. The Tongshi gold field includes three basic geological units according to **Figure 15**, (a) a concealed basement (**I**), with positive Bouguer anomaly in **Figure 14**, for which positive Bouguer anomaly at northeast margin might be the reflection of Archean metamorphic basement swells; (b) Mesozoic volcanic-sedimentary basin (**II**) extending in NW direction exhibits an obvious negative gravity anomaly; and (c) the Tongshi intrusive complex (**III**) with negative Bouguer anomaly intruded into the concealed basement. According to the extracted Bouguer anomaly features, Unit **III** can be subdivided into subunits **IIIa** and **IIIb**. The circular positive gravity anomaly (**IIIb**) around an obvious negative anomalous center (**IIIa**) might reflect the spatial distribution of the alteration and mineralization halo of the Tongshi intrusive complex, which might represent the contact metasomatic zone that developed between the Tongshi intrusive complex and its host rocks. The spatial distributions of various types of gold deposits are controlled by Unit III (**Figure 15**).

In conclusion, BEMD was successfully applied to the decomposition of gravity data for two study areas. The BEMD method is not only suitable for analysis of gravity data, but also can be extended for analysis of the scattered data like magnetic data and exploration geochemical data.

Acknowledgements

We thank Maja Bozicevic from InTechOpen for invitation. This research was jointly funded by the NSFC Program (41672329, 41272365) and the Chinese Program for probing into deep Earth (2016YFC0600509). We also thank both Yunnan Geological Survey and Yanzhou Geology & Mineral Exploration Institute of Shandong province for providing original gravity data for this research.

Author details

Yongqing Chen[1*], Binbin Zhao[2], Jingnning Huang[3] and Lina Zhang[4]

*Address all correspondence to: yqchen@cugb.edu.cn

1 China University of Geosciences (Beijing), Beijing, China

2 Sinomine Resource Exploration Co., LTD., Beijing, China

3 The Geological Museum of China, Beijing, China

4 Reservoir Evaluation Center, CNPC Logging Co., LTD., Xi'an, China

References

[1] Huang NE, Shen Z, Long SR, Wu MC, Shi HH, Zheng Q, Yen NC, Tung CC, Liu HH. Mint: The empirical mode decomposition and the Hilbert spectrum for nonlinear and non-stationary time series analysis. Proceedings of the Royal Society of London A. 1998;**454**:903-995

[2] Nunes JC, Bouaoune Y, Delechelle E, Niang O, Bunel P. Mint: Image analysis by bi-dimensional empirical mode decomposition. Image and Vision Computing. 2003;**21**:1019-1026

[3] Nunes JC, Guyot S, Delechelle E. Mint: Texture analysis based on local analysis of the bidimensional empirical mode decomposition. Machine Vision and Applications. 2005;**16**: 177-188

[4] Huang NE, Shen SSP. Hilbert-Huang Transform and Its Applications. Singapore: World Scientific Publishing Co. Pte. Ltd.; 2005. pp. 311

[5] Freire SLM, Ulrych TJ. Mint: Application of singular value decomposition to vertical seismic profiling. Geophysics. 1988;**53**:778-785

[6] Cheng YB, Mao JW, Spandler C. Mint: Petrogenesis and geodynamic implications of the Gejiu igneous complex in the western Cathaysia block, South China. Lithos. 2013;**175-176**:213-229

[7] Party G. Geology of Tin Deposit in Gejiu Area. Beijing: Metallurgical Industry Publishing House; 1984. p. 50-90

[8] Qin DX, Li YS. Studies on the Geology of the Gejiu Sn-Cu Deposit. Beijing: Science Press; 2008

[9] Cheng YB, Mao JW. Mint: Age and geochemistry of granites in Gejiu area, Yunnan province, SW China: Constraints on their petrogenesis and tectonic setting. Lithos. 2010;**120**:258-276

[10] Cheng YB, Mao JW, Rusk B, Yang ZX. Mint: Geology and genesis of Kafang Cu–Sn deposit, Gejiu field, SW China. Ore Geology Reviews. 2012;**48**:180-196

[11] Cheng YB, Mao JW, Chang ZS, Pirajno F. Mint: The origin of the world class tin-polymetallic deposits in the Gejiu field, SW China: Constraints from metal zoning characteristics and ^{40}Ar–^{39}Ar geochronology. Ore Geology Reviews. 2013;**53**:50-62

[12] Xiong GC, Shi ST. Mint: Physico-geologic model of the Gejiu tin district and its application. Geological Review. 1994;**40**:19-27 (in Chinese with English abstract)

[13] Chen YQ, Zhang LN, Zhao BB. Mint: Application of singular value decomposition (SVD) in extraction of gravity components indicating the deeply and shallowly buried granitic complex associated with tin polymetallic mineralization in the Gejiu tin ore field, Southwestern China. Journal of Applied Geophysics. 2015;**123**:63-70

[14] Cheng YB, Mao JW, Chen XL. Mint: LA-ICP-MS Zircon U-Pb Dating of the Bozhushan Granite in Southeastern Yunnan Province and Its Significance. Journal of Jilin University (Earth Science Edition). 2010;**40**:869-878 (in Chinese with English abstract)

[15] Liu JS, Zhang HP, Ouyang YF, Zhang CH. Mint: Bainiuchang super-large silver-polymetallic ore deposit related to granitic magmatism in Mengz, Yunnani. J. Cent. South Univ. Technol. 2007;**14**:568-574

[16] Hou WS, Yang ZJ, Zhou YZ, Zhang LP, Wu WL. Mint: Extracting magnetic anomalies based on an improved BEMD method: A case study in the Pangxidong area, South China. Computers & Geosciences. 2012;**48**:1-8

[17] Wang XF, Metcalfe I, Jian P, He LQ, Wang CS. Mint: The Jinshajiang–Ailaoshan Suture zone, China: Tectonostratigraphy, age and evolution. Journal of Asian Earth Sciences. 2000:675-690

[18] Harrison TM, Chen WJ, Leloup PH. Mint: An early Miocene transition in deformation regime within the Red River fault zone, Yunnan, and its significance for indo-Asia tectonic. Journal of Geophysics Research. 1992;**97**:7159-7182

[19] Leloup PH, Lacassin R, Tapponnier P. Mint: The Ailaoshan-Red River shear zone (Yunnan, China). Tertiary transform boundary of Indochina, Tectonophysics. 1995;**251**:3-83

[20] Gilley LD, Harrison TM, Leloup PH, Ryerson FJ, Lovera OM, Wang JH. Mint: Direct dating of left-lateral deformation along the Red River shear zone, China and Vietnam. Journal of Geophysical Research. 2003;**108**:14-21

[21] Zhang JJ, Zhong DL, Sang HQ. Mint: Structural and geochronological evidence for multiple episodes of tertiary deformation along the Ailaoshan-Red River shear zone, southeastern Asia, since the Paleocene. Acta Geologica Sinica. 2006;**80**:79-96

[22] Liu JL, Chen XY, Wu WB, Tang Y, Tran MD, Nguyen QL, Zhang ZC, Zhao ZD. Mint: New tectono-geochronological constraints on timing of shearing along the Ailao Shan-Red River shear zone: Implications for genesis of Ailao Shan gold mineralization. Journal of Asian Earth Sciences. 2015;**103**:70-86

[23] Tran MD, Liu JL, Nguyen QL, Chen Y, Tang Song ZJ, Zhang ZC, Zhao ZD. Mint: Cenozoichigh-K alkaline magmatism and associated Cu–Mo–Au mineralization in the Jinping–Fan Si Panregion, southeastern Ailao Shan–Red River shear zone, southwestern China–northwestern Vietnam. Journal of Asian Earth Sciences. 2014;**79**:858-872

[24] Kane MF, Godson RH. Mint: Features of a pair of long-wavelength (>250km) and short-wavelength (<250km) Bouguer gravity maps of the United States. In: Hinze WJ, editor. The Utility of Regional Gravity and Magnetic Anomaly Maps. Society of Exploration Geophysicists; 1985. p. 46-61

[25] Zeng HL, Zhang QH, Li YS, Liu J. Mint: Crustal structure inferred from gravity anomalies in South China. Tectonophysics. 1997;**283**:189-203

[26] Shah AK, Bedrosian PA, Anderson ED, Kelley KD, Lang J. Mint: Integrated geophysical imaging of a concealed mineral deposit: A case study of the world-class pebble porphyry deposit in southwestern Alaska. Geophysics. 2013;**78**:317-328

[27] Chen YQ, Xia QL, Liu HG. Mint: Delineation of potential mineral resources region based on geo-anomaly unit. Journal of Earth Science. 2000;**11**(2):158-163 (in Chinese)

[28] Chen YQ, Zhao PD, Chen JG, Liu JP. Mint: Application of the geo-anomaly unit concept in quantitative delineation and assessment of gold ore targets in western Shandong uplift terrain, Eastern China. Natural Resources Research. 2001;**10**(1):35-49

[29] Lin JQ, Tan DJ. Mint: 40Ar/39Ar ages of Mesozoic igneous activities in western Shandong. Acta Petrologica et Mineralogica. 1996;**15**(3):213-220 (in Chinese)

[30] Hu HB, Mao JW, Niu SY, Li MW, Chai FM, Li YF, Liu T. Mint: Study on ore- forming fluids of the Guilaizhuang gold deposits in Pingyi, Western Shandong. Journal of Mineralogy and Petrology. 2005;**29**(1):38-44 (in Chinese)

[31] Wang SC, Liu YQ, Yi PH, Zhang YQ, Yang YH, Yang DL, Huang TL, Fan JZ, Ye SX, Dong T, Chen YQ, Ye SS. Mint: Gold Deposits and Synthetic Information Metallogenic Prognosis for Gold Deposit Concentrated Areas. Beijing, China: Geological Publishing House; 2003245pp (in Chinese)

Chapter 5

Identification of Gravity Lineaments for Water Resources in the Crystalline Massif of Hoggar (South of Algeria)

Abdeslam Abtout, Hassina Boukerbout,
Boualem Bouyahiaoui and Farida Boukercha

Additional information is available at the end of the chapter

http://dx.doi.org/10.5772/intechopen.71204

Abstract

The Hoggar is a rock mountain located in the South of Algeria. It is a crystalline massif characterized by granitic substratum with a weak sedimentary cover. The pluviometry is low in this region and is characterized by possibilities of floods of big capacities of water. The infiltration of these important quantities of water remains low because of the insufficiency of systems of retains. The problem in this case is locating the water in weathered zones, beyond 50 m, above granites. The weathered areas could be sometimes aquifer at their base. So, this can be interesting if they rise above intensely fractured rocks and/or suitable geometry of the substratum. The fractured rocks and the suitable geometry of the substratum could be linked to natural reservoirs. The perfect case is that this structure is covered by a rather important thickness of silts to let the infiltration of the water. So, to identify the various juxtaposed structures with different densities and delineate gravity lineaments, faults, and cavities, the gravimetric method was preferred. The aim of this work is integrating all geometric and gravimetric observations, models, and approaches so as to provide consistent and reliable information for making decision regarding the location of drilling.

Keywords: gravity, anomalies, structures, faults, granitic substratum, silts, filtering, modeling, continuous wavelet transform, drilling, water

1. Introduction

The aim of the gravimetric prospecting is to measure the gravity field variations and to interpret the anomalies caused by lateral homogeneities of densities and crust thickness.

The interpretation target is to identify the anomalies' causative structures. It means shaping the structures by their geometrical form, depth, dimension, and so on and their physical properties such as density contrast, which could be answers to the posed problem.

The gravimetric anomalies shape reveals geological bodies' geometry. The structures may be represented by simple geometrical forms such as spheres for saliferous dome, heap mineralized, underground cavities, or as cylinders for folds, dykes, ore bodies, or half-plans for substratum fracturing, faults, flexures, thrusting, subduction zones, and so on.

The relative measures of gravity aim at determining the variations of the gravity g from a station or measure point to the other one. The variation of g for each station is measured in relation to the value in the base station. These measures are achieved with gravimeters. The basic gravimeter is a pendulum in which a mass m is suspended from a spring. The variation of g between two sites causes an extension of the spring. As the extension of the spring is proportional to the gravity, the variations of g between the base station and the successive stations can be measured by determining the value of g at each station.

The gravity upon the earth surface varies according to five factors: the latitude, the altitude, the surrounding relief, the astronomical effects, terrestrial tide, and the density of rocks in the crust and upper mantle.

The variations of g related to the first four factors are attributable to external causes. Only the variations of g reflecting density variations of the crust structures are considered, such as those due to the internal causes and represent an interest for gravimetric prospecting. Hence, the measured values of g in the field must be corrected from the external effects so that the gravity anomalies would be highlighted.

The Bouguer anomaly map is the basic document for interpreting gravimetric data. It is the difference between the measured and corrected gravity for each measured point (Faye, Bouguer, and terrain corrections) and the theoretical value computed for the same point on the ellipsoid. The Bouguer anomaly map enhances the deep and extended structures. The short wavelength anomalies on the Bouguer anomaly map correspond to density variations caused by structures in the crust such as substratum fracturing, saliferous dome, underground cavities, ore zone, and so on.

The Bouguer anomaly is the sum of contributions of each formation in the basement. As we are interested to individual contribution, filtering is then necessary to detect the bodies of geological interest. The advantage in water prospecting is that the interesting structures display very clear gravimetric characteristics that enable the choice of the anomalies to keep. After filtering the anomalies (gradient, continuation…), the localization of the causative structures of the anomalies is determined by the continuous wavelet transform method.

The gravity prospection concerns five terrains in the Tamanrasset region, positioned at coordinates: 05°31′–05°38′ of longitude and 22°47′–22°52′ of latitude. The studied area is located in the Hoggar (**Figure 1**) which is a crystalline massif. The prospected zone is characterized by a domain of metamorphic Precambrian rocks, intruded in Pharusian by granites and covered in some sites by volcanic sewage (**Figure 2**). The different tectonic phases inferred a complex structure characterized by many tectonic accidents which affect the substratum in most of the cases. The climate of this region is classified as arid with low rainfalls and high evaporation. Many oueds cross the area coming from the Mount of Atakor. The area has: a granitic substratum which is impermeable, an altered or weathered substratum in favor of an accumulation of water; silts acting as main reservoir and finally fractured zones and contacts allowing water flow. This work leads to investigate and to identify buried structures such as contacts and faults which could be traps, where water can circulate, determining the thickness of alluvial curvature, so that these information could help for making decision regarding the location of drilling. To achieve that, the gravity data are processed with different methods such as gradients, upward continuation, and continuous wavelet transform; this can lead to an updated structural map for the studied area based on Bouguer gravity anomaly data. The most important information, in terms of contrast density structures at nearsurface or depth, will be very useful for making decision regarding the location of drilling.

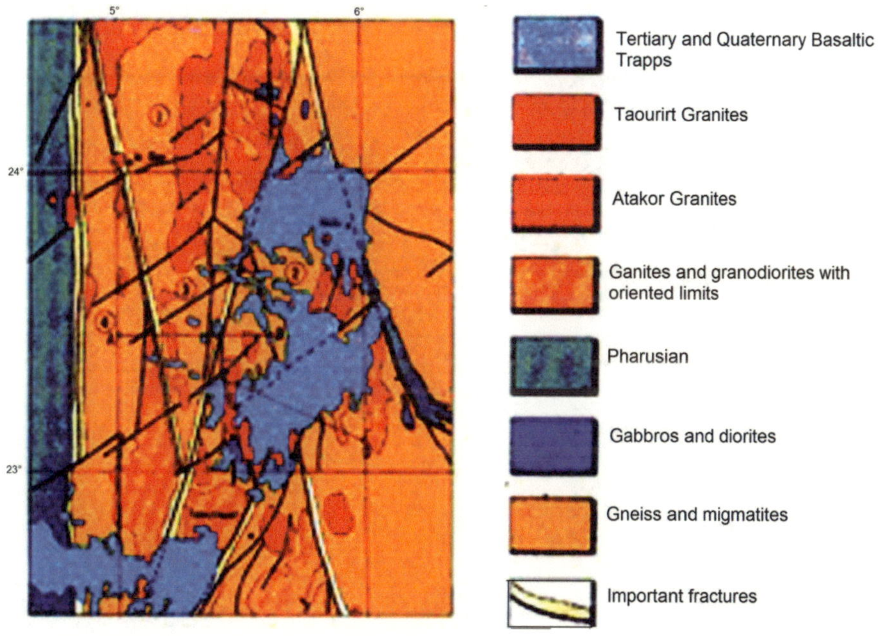

Figure 1. Geological map of the Hoggar.

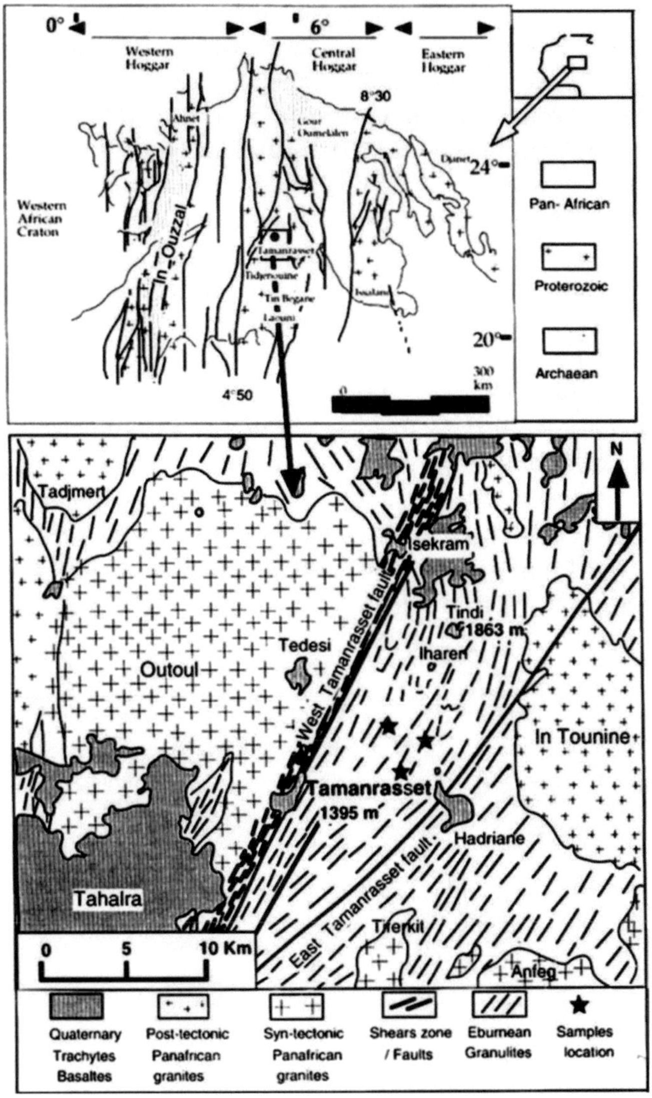

Figure 2. Geological map of the surveyed area.

2. Gravity prospecting

The gravity survey is achieved by the use of two different micro gravimeters "Lacoste Romberg" and "Scintrex CG3" (**Figure 3**). These kinds of gravimeters are suitable for such local studies because of their good instrumental precision of 1 μgal and meet the most stringent accuracy requirements. The devices are endowed with an electronic system of reset, with an automatic controller of bubbles level and, a system of storage and data pre-processing. All this confers them a precision of effective measure lower than 5 μgal. To tie the gravimetric network, the

Figure 3. The gravimeter Lacoste and Romberg.

international gravimetric reference station of Tamanrasset is used. From this reference station, secondary gravimetric basis located in the center of each surveyed area are realized. Otherwise, the sampling measurement points is 20 m, so that the density contrast of structures located nearsurface or at low depth could be clearly highlighted. For every surveyed zone, an extension of the measures along profiles beyond the limits of the zone was completed in order to avoid edge effects and artifacts in processing. There are more than 2500 measurement points for the five surveyed areas (**Figure 4**). The altitude at every measure point performed with an automatic level type WILD "SOKKIA" B20 with compensator is used. The precision of the device is about 0.5 mm or 2 mm/km. The altitudes of prospected areas are determined with a precision of 2 cm (**Figure 5**). The coordinates limiting the surveyed areas are determined by single-frequency GPS (Garmin GPS 12) and the meshing of measurement points is made with the use of theodolite (Zeïss 020 B).

The Bouguer anomaly is performed as follow:

$$A_b = G_m - G_{th} + A_f - P + T$$

where A_b is the Bouguer anomaly, G_m is the measured gravity, G_{th} is the gravity computed on the reference ellipsoid or latitude correction, A_f is the "free air" or Faye correction, P is Bouguer correction, and T is terrain correction.

As the investigated areas have limited dimensions, the latitude correction is supposed constant, hence only the Bouguer correction is applied to the observed gravity. This correction depends on density and altitude and is defined as follow:

$$A_f - P = (0.3086 - 0.0419) \cdot h$$

in mGal/m. The density d = 2.7 is determined by the method of Nettleton. As the altitudes precision is 2 cm, the corrections are determined with precision of 4 μgal and the Bouguer anomaly values are determined with precision of 7 μgal.

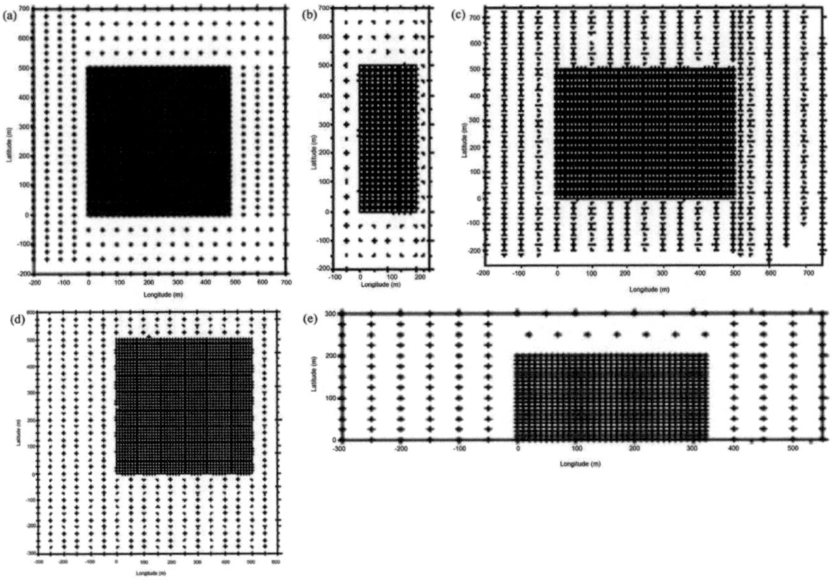

Figure 4. (a) Location of gravity measurements points of terrain 1 with the extensions. (b) Location of gravity measurements points of terrain 2 with the extensions. (c) Location of gravity measurements points of terrain 3 with the extensions. (d) Location of gravity measurements points of terrain 4 with the extensions. (e) Location of gravity measurements points of terrain 5 with the extensions.

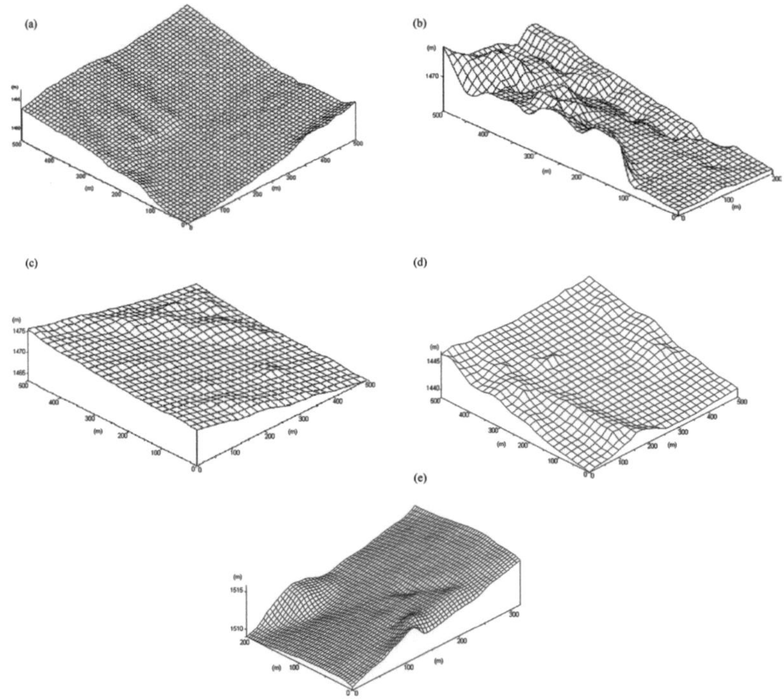

Figure 5. (a) Altitudes map of terrain 1. (b) Altitudes map of terrain 2. (c) Altitudes map of terrain 3. (d) Altitudes map of terrain 4. (e) Altitudes map of terrain 5.

The Bouguer anomaly maps are drawn on a regular grid with the use of "inverse distance" interpolation method and an equidistance of levels of 100 μgal.

3. Data processing

Once the Bouguer anomaly maps are established for the five zones, all processing methods are applied on the Bouguer anomaly data to extract the useful information. In what is going to follow, we choose to show the results obtained for one surveyed area (Terrain 4), knowing that all processing methods were applied for the four other areas.

The studied area which corresponds to terrain 4 or site n°4 in **Figure 6**, is located at 2 km in the North of Tamanrasset. The gravity survey covers a surface of 500 × 500 m, with equidistance of 20 m (**Figure 4d**). The map of the Bouguer anomaly is the basic document for gravimetric interpretation and this map is the sum of different measured effects: superficial, intermediate, and deep, in vertical direction. Thus, to separate all these effects, it is necessary to apply different filters. As we are interested in local anomalies, the first processing method used consists in eliminating the regional effect from the Bouguer anomaly map by the use of the polynomial method at different orders [1]. For a better identification of gravity lineaments, different methods are used, such as the vertical and directional gradients [2] which are used to highlight the short wavelength anomalies. On the other hand, the upward continuation at different altitudes [3] method enhances the long wavelength anomalies by attenuating superficial anomalies and strengthening deep ones [4–6]. To localize the anomalies causative structures at surface and depth, the continuous wavelet transform method is used [7–10].

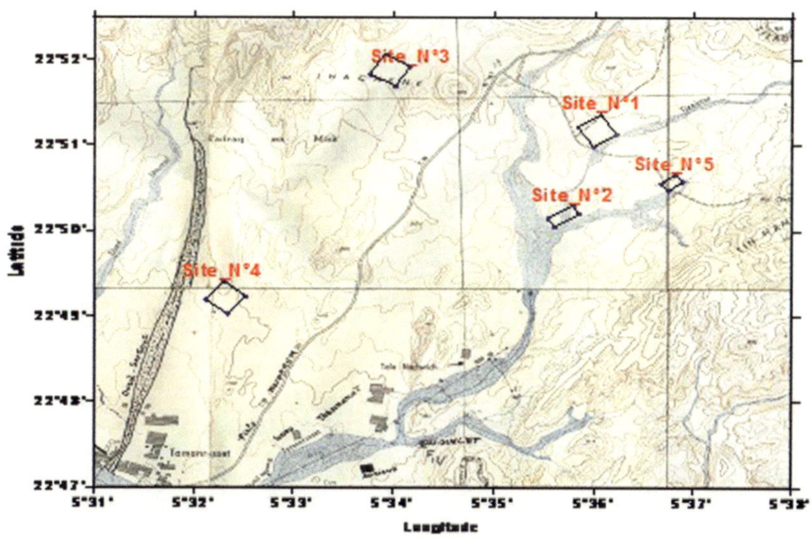

Figure 6. Location of the five surveyed areas in Tamanrasset region.

4. Interpretation and results

4.1. Description of processed maps

The Bouguer anomaly map (**Figure 7**) shows a series of positive and negative anomalies elongated in the NE-SW direction. The values of the Bouguer anomalies are ranged between −110.650 and −111.500 mGal. The examination of this map suggests that the regional effect could be plane or paraboloid. The upward continued map of the Bouguer anomaly at an altitude of 150 m (**Figure 8**) shows that the regional effect is a paraboloïd. **Figures 9a** and **b** correspond to the regional anomalies and, **Figures 10a** and **b** show the residual anomalies maps of order 1 and order 2, obtained by retrieving polynomials of order 1 and 2. The residual anomaly maps show the same trends as the Bouguer anomaly map, principally a series of negative anomalies in the center of the area bordered by positive anomalies.

The maps of horizontal gradient in x direction (**Figure 11a**), shows succession of anomalies in NNW-SSE direction, while the derivative in y direction (**Figure 11b**) illustrates an important gradient in the E-W direction. The vertical gradient map (**Figure 11c**) points out gravity lineaments in NNW-SSE and E-W directions.

The long wavelength anomalies are highlighted by the use of upward continuation at different altitudes (**Figure 12**). At greater altitudes than 150 m, the surveyed area is characterized by a positive anomaly and suggests that the structure responsible of this anomaly is relatively

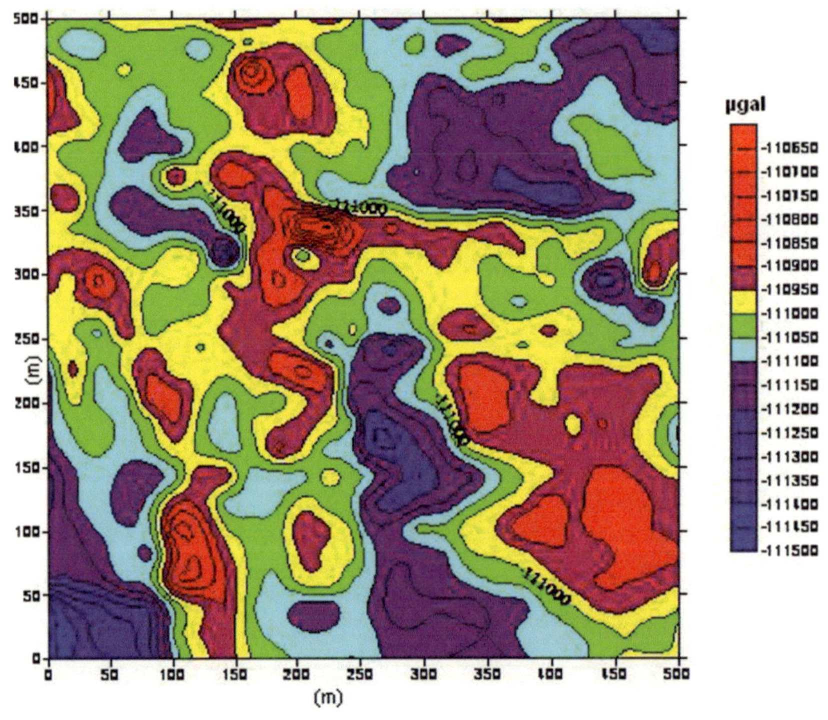

Figure 7. The Bouguer anomaly map of terrain 4.

http://dx.doi.org/10.5772/intechopen.71204

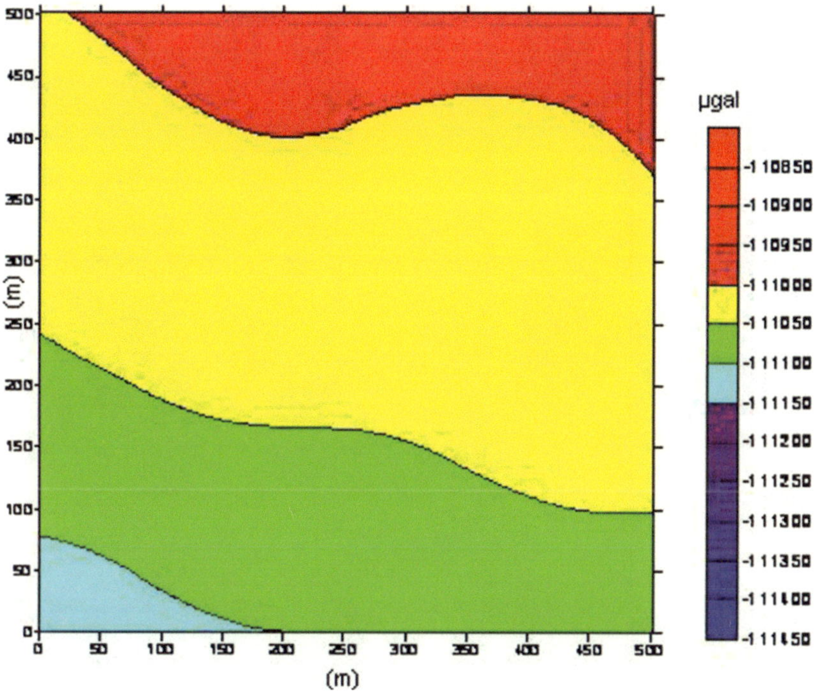

Figure 8. Upward continuation at 150 m of the Bouguer anomaly map of terrain 4.

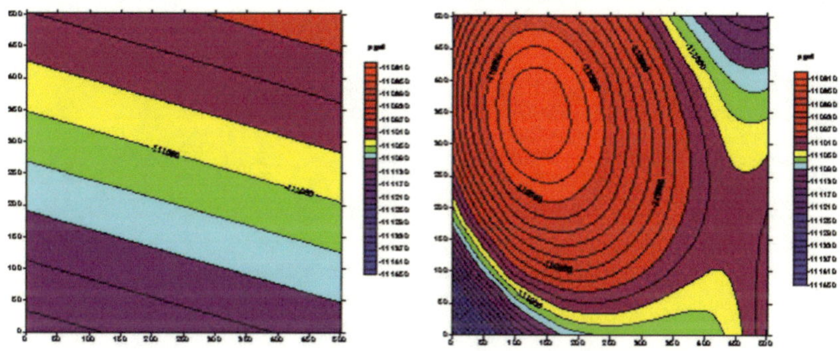

Figure 9. Regional anomalies maps of order 1 (left) and order 2 (right) of terrain 4.

deeply rooted, more than 150 m of depth. Between the altitudes of 10 and 50 m, the positive anomaly is formed by two different parts, crossed with a negative anomaly in the central part of the area along an N-S axis.

4.2. Identification and localization of anomalies causative bodies

The continuous wavelet transform method is used to identify and localize structures responsible of the anomalies both at surface and in depth. The method is applied on five profiles perpendicular to residual anomalies directions (**Figure 10**). The results obtained are displayed

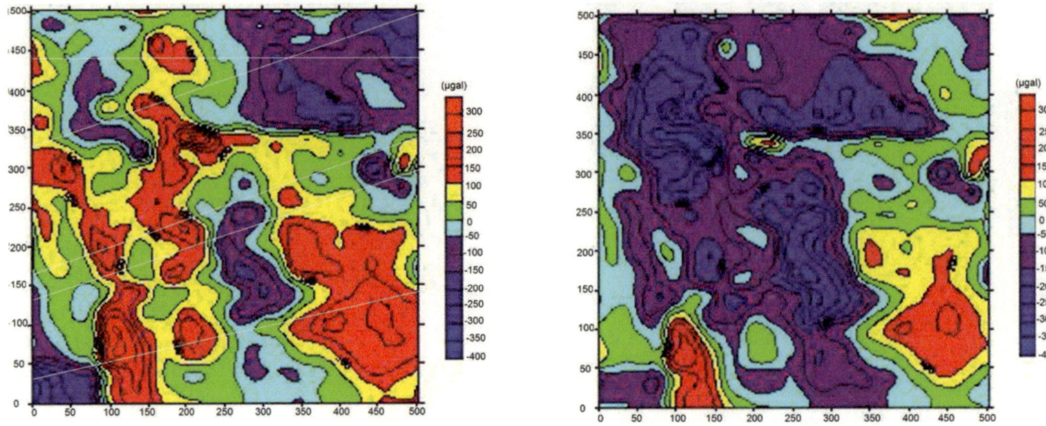

Figure 10. Residual anomalies maps of order 1 (left) and order 2 (right) of terrain 4. In white lines the processed profiles with the continuous wavelet transform in the Section 4.2.

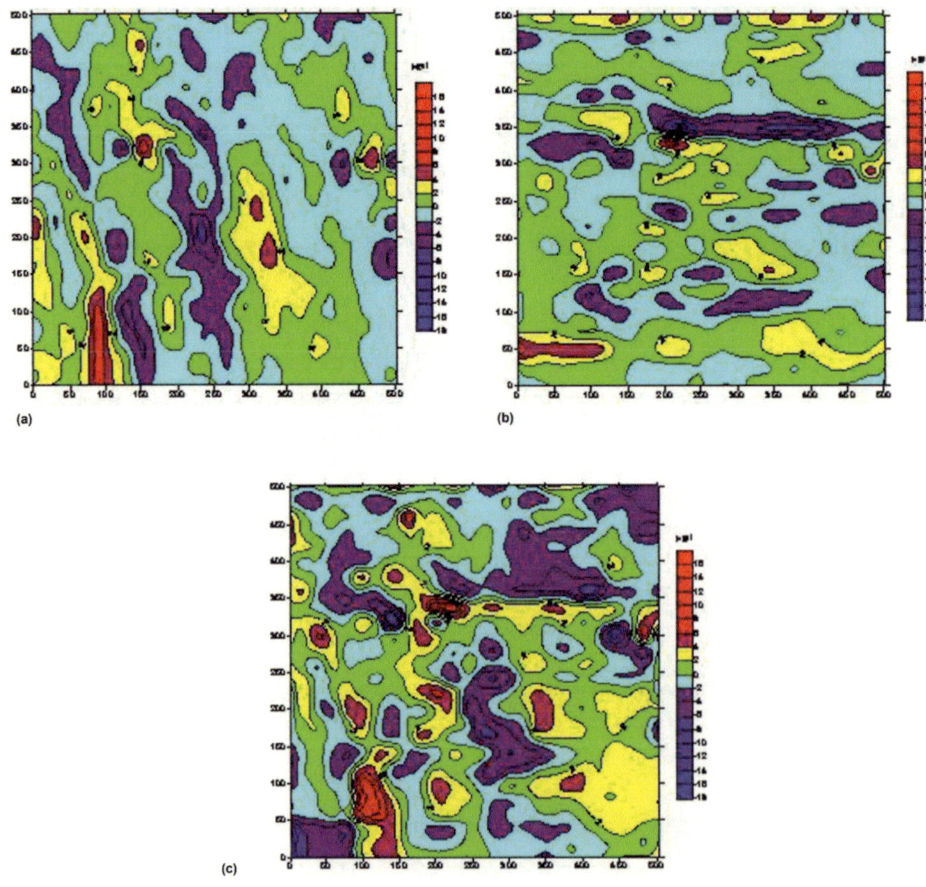

Figure 11. Gradients maps of terrain 4. (a) Horizontal gradient in x direction, (b) in y direction, and (c) vertical gradient.

http://dx.doi.org/10.5772/intechopen.71204

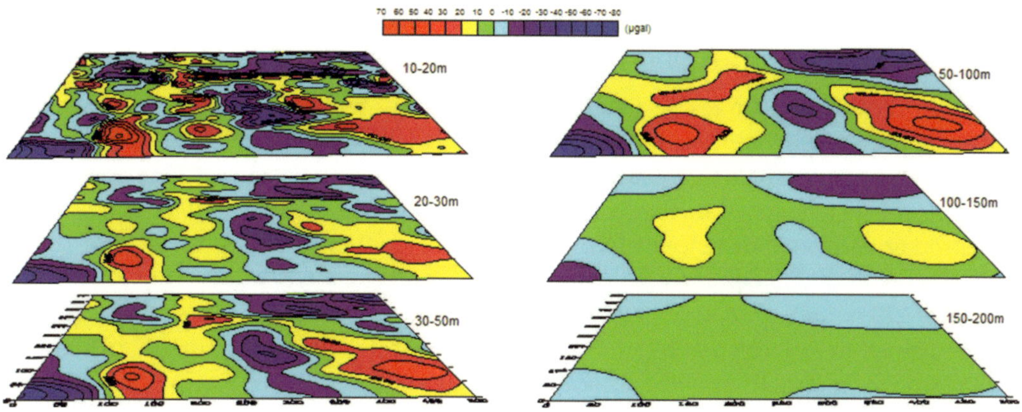

Figure 12. Upward continued maps differences of the Bouguer anomaly of terrain 4 at different altitudes (10–20 m, 20–30 m, 30–50 m, 50–100 m, 100–150 m, and 150–200 m).

in **Figure 13**. The figure in the top represents the signal corresponding to the gravity anomaly in mGal according to horizontal distance in m. The figure in the middle corresponds to the map of the modulus of the coefficients of the wavelet transform according to dilations in m and horizontal distances in m and, the figure in the bottom corresponds to the map of the maximum entropy according to horizontal distance in m and the depth of identified structures in m.

The **Figure 13a** shows the results obtained for the gravity profile 1 of the studied area. The intensity of the anomaly varies from −150 to 300 mGal. The map of the wavelet transform (middle) displays the signature of many contacts or faults. The first contact is identified in the entropy map, at horizontal distance of 100 m and at depth of 100 m. Between the distances of 120 and 140 m, a structure is identified at 40 m of depth. A second contact is identified at horizontal distance of 200 m and at depth of 100 m. At a distance of 220 m, the depth of the identified structure is 70 m. Between the distances of 221 and 230 m, the depth of identified structure is 40 m. A third contact is identified at horizontal distance of 260 m and at depth greater than 100 m. Between the distances 280 and 300 m, the identified structure is located at a depth of 65 m. The last identified contact along this profile is localized at a distance of 400 m and at depth greater than 100 m. The last identified structure is located at distances 480 and 520 m and at depth of 80 m.

The **Figure 13b** shows the results obtained for the gravity profile 2 of the studied area. The intensity of the anomaly varies from −180 to 150 mGal. The map of the wavelet transform (middle) displays the signature of many contacts. The first one is identified in the entropy map, at horizontal distance of 100 m and at depth of 80 m. A structure is identified at 30 m of depth and horizontal distance of 120 m. A second contact is identified at horizontal distance of 140 m and at depth of 100 m. At a distance of 240 m, the depth of the identified structure is 30 m. A third contact is identified at horizontal distance of 260 m and at depth of 35 m. Between the horizontal distances of 280 and 300 m, the identified structure is located at a depth of 70 m while between the distances of 320 and 340 m, its depth reaches 65 m. Another identified contact along this profile is localized at a distance of 460 m and at depth of 70 m. The last identified structure is located between distances 480 and 490 m and at depth of 40 m.

The **Figure 13c** displays the results obtained for the gravity profile 3 of the surveyed area. The intensity of the anomaly varies from −100 to 150 mGal. Many contacts and structures are identified along this profile. The first contact is identified at horizontal distance of 20 m and at depth of 30 m. At the distance of 30 m, the structure is identified at 12 m of depth. The second contact is localized at horizontal distance of 60 m and at depth of 90 m. Between distances of 180 and 210 m, the depth of the identified structure is 80 m. Another contact is identified at horizontal distance of 220 m and at depth greater than 100 m. Between the distances 260 and 285 m, the identified structure is located at a depth of 85 m, while at the horizontal distance of 300 m, its depth reaches 20 m. Another contact is localized at a distance of 320 m and at depth of 70 m. The identified structure is located at horizontal distance 340 m and at depth of 20 m and another contact is identified at 360 m with a depth of 55 m. At the distance of 400 m, the depth of the structure reaches the 10 m and the last contact is identified at a depth of 70 m and horizontal distance of 480 m.

The **Figure 13d** displays the results obtained for the gravity profile 4 of the surveyed area. The intensity of the anomaly varies from −220 to 150 mGal. The first structure is identified between horizontal distances of 10 and 30 m, at depth of 10 m and reaches the 40 m of depth at distance of 40 m. An identified contact is located at distance of 60 m and the depth of 80 m. Between the distances of 80 and 100 m, the depth of the structure reaches 25 m. The second contact is localized at horizontal distance of 140 m and at depth exceeding 100 m. Another contact is identified at a distance of 320 m and the depth of 90 m. Between distances of 330 and 350 m, the depth of the identified structure is about 50 m.

The **Figure 13e** shows the results obtained for the gravity profile 5. The intensity of the anomaly varies from −200 to 150 mGal. The first identified structure is localized between the distances 10 and 30 m at 10 m of depth, further at the distance of 100 m its depth reaches 20 m. Two contacts are identified at horizontal distances of 140 and 260 m respectively deeper than 100 m.

By stacking the results of the continuous wavelet transform applied on more profiles extracted from gravity anomalies map of this area, we establish the depth map for this zone. The gravity anomalies and depth maps are overlaid (**Figure 14**), two weathered zones elongated in the NW-SE appear on this map, and these weathered zones are staggered relative to each other by a fault with NE-SW direction. Another weathered zone is identified in the NE part of the area and is fairly deeply rooted. The depth of these weathered zones is about 120 m.

4.3. Discussion

A compilation of identified structures for the five surveyed areas is shown in **Figure 14**.

For the first terrain (**Figure 14a**), the granitic substratum rises in the NW and SW parts of the terrain 1 and it is crossed in its central part by a fault. It is covered by two weathered zones elongated in the N-S direction and located at longitudes of 150 and 200 m, respectively. The

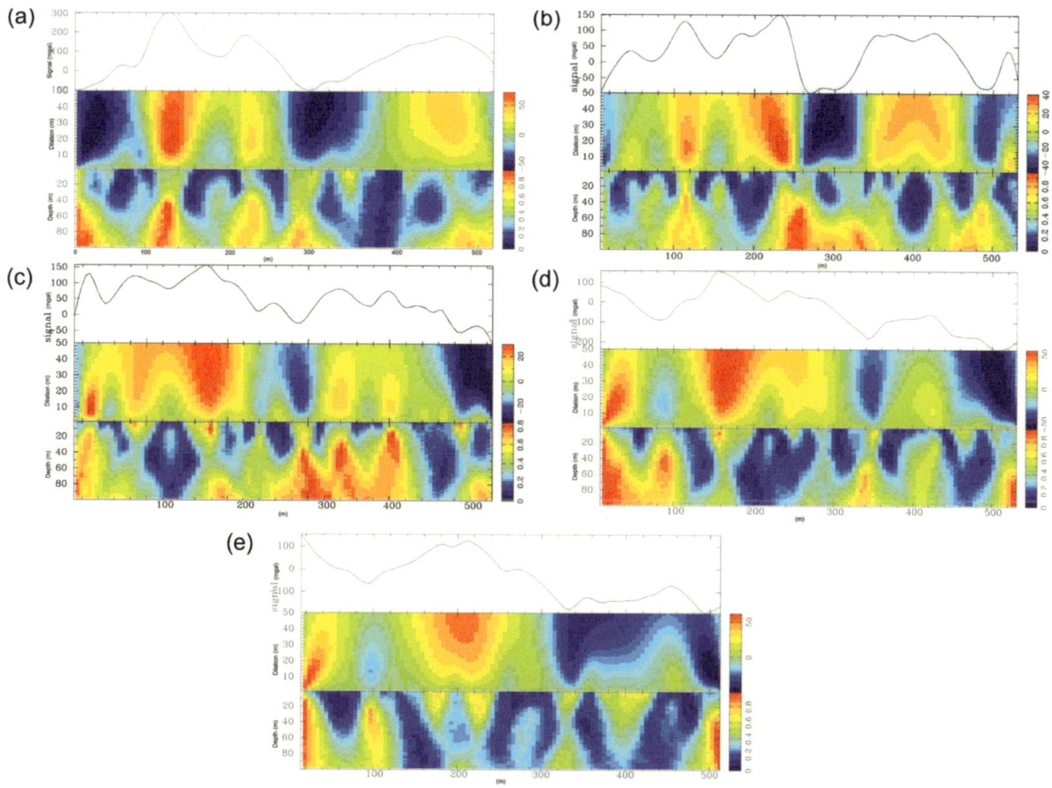

Figure 13. (a) Localization of gravity anomalies causative bodies along the profile 1 of terrain 4. (b) Localization of gravity anomalies causative bodies along the profile 2 of terrain 4. (c) Localization of gravity anomalies causative bodies along the profile 3 of terrain 4. (d) Localization of gravity anomalies causative bodies along the profile 4 of terrain 4. (e) Localization of gravity anomalies causative bodies along the profile 5 of terrain 4.

depth of these weathered zones reaches 75 m and at this depth the substratum appears highly fractured. This map shows the deepest sources are located at longitudes of 70 and 150 m with a maximum at latitude of 270 m.

The **Figure 14b** shows that the granitic substratum is very deep in the terrain 2 and rises in the NW and SE parts of the area. In the central part of the terrain, the substratum remains very deep. In the NE and southernmost parts of the surveyed terrain, the substratum is covered by weathered zones. The structures depth is very important in the northern part of the area which is intensely fractured. In the terrain 3 (**Figure 14c**), the granitic substratum rises in the eastern part of the area where it is crossed by a fault of NW-SE direction and, in the western part of the area where it is less deep. In the central part of the area, some weathered zones of low thickness cover the granitic substratum; here, the depths do not exceed the 10 m. In the southern part of the area, the depths become very important but it is the limit of the terrain 3. In the terrain 4 (**Figure 14d**), the granitic substratum is deep. Two weathered zones in NW-SE direction appear and are shifted with respect to each other by a fault in NE-SW direction. Another weathered deep zone is located in the NE part of the area. The weathered zones

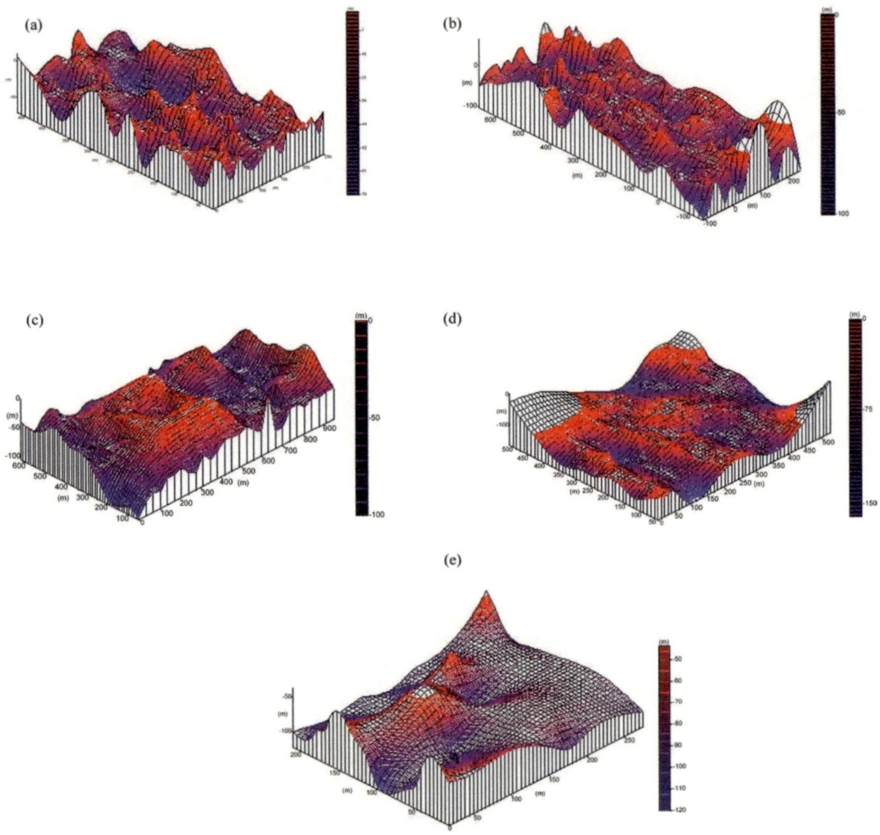

Figure 14. (a) Structures depths map of terrain 1. (b) Structures depths map of terrain 2. (c) Structures depths map of terrain 3. (d) Structures depths map of terrain 4. (e) Structures depths map of terrain 5.

depth is about 120 m. The **Figure 14e** shows a weathered zone in the central part of the terrain 5 which is crossed by a series of faults with NNE-SSW direction. The weathered zones depth reaches 80 m.

5. Conclusion

The gravity survey conducted in the Tamanrasset region and covering five areas lead to map gravity lineaments. By the use of the different processing methods (gradients, upward continuation, wavelet transform) applied on gravity anomalies data, the faults, contacts, granitic substratum, and weathered zones are identified and localized at near surface and at depth. On the basis of this work and achievements, drillings were set up in every surveyed area. There were three areas where the water was found and allowed to feed the city of Tamanrasset. Ten years later, a definitive solution to the problem of the water supply for the Tamanrasset city was found; the water is returned by the city of In Salah toward the city of Tamanrasset via a 700 km water main.

Acknowledgements

This work is supported by the Centre de Recherche en astronomie, Astrophysique & Géophysique-Observatoire d'Alger. We are grateful to Mr. A. Yelles, Ms. M. Bozicevic and Ms. R. Rovan. We gratefully acknowledge the editor and reviewers for their constructive remarks.

Author details

Abdeslam Abtout, Hassina Boukerbout*, Boualem Bouyahiaoui and Farida Boukercha

*Address all correspondence to: s.boukerbout@craag.dz

Centre de Recherche en Astronomie, Astrophysique & Géophysique-Observatoire d'Alger, Alger, Algeria

References

[1] Oldham CW, Sutherland DB. Orthogonal polynomials: Their use in estimating the regional effect. Geophysics. 1955;**20**:295-306

[2] Baranov V. Calcul du gradient vertical du champ de gravité ou du champ magne'tique mesure' a` la surface du sol. Geophysical Prospecting. 1953;**1**(3):171-191

[3] Bhattacharyya BK, Chan KC. Reduction of magnetic and gravity data on an arbitrary surface acquired in a region of high topographic relief. Geophysics. 1977;**42**(7):1411-1430

[4] Gibert D, Galdeano A. A computer program to perform transformations of gravimetric and aeromagnetic survey. Computers & Geosciences. 1985;**11**:553-588

[5] Abtout A, Boukerbout H, Bouyahiaoui B, Gibert D. Gravimetric evidences of active faults and underground structure of the Cheliff seismogenic basin (Algeria). Journal of African Earth Sciences. 2014;**99**:363-373. DOI: 10.1016/j.jafrearsci.2014.02.011

[6] Bouyahiaoui B, Abtout A, Hamai L, Boukerbout H, Djellit H, Bougchiche S, Bendali M, Bouabdallah H. Structural architecture of the hydrothermal system from geophysical data in Hammam Bouhadjar area (Northwest of Algeria). Pure and Applied Geophysics. 2017;**174**(3):1471-1488. DOI: 10.1007/s00024-017-1479-0

[7] Moreau F, Gibert D, Holschneider M, Saracco G. Wavelet analysis of potential fields. Inverse Problems. 1997;**13**:165-178

[8] Hornby P, Boschetti F, Horovitz FG. Analysis of potential field data in the waveletdomain. Geophysical Journal International. 1999;**137**:175-196

[9] Boukerbout H, Gibert D, Sailhac P. Identification of sources of potential fields with the continuous wavelet transform: Application to VLF data. Geophysical Research Letters. 2003;**30**(8):1427. DOI: 10.1029/2003GL016884.

[10] Sailhac P, Gibert D, Boukerbout H. The theory of the continuous wavelet transform in the interpretation of potential fields: A review. Geophysical Prospecting. 2009;**57**(4):517-525

Chapter 6

Gravity Application for Delineating Subsurface Structures at Different Localities in Egypt

Sultan Awad Sultan Araffa

Additional information is available at the end of the chapter

http://dx.doi.org/10.5772/intechopen.71492

Abstract

Gravity method is important tool for delineating structural elements which dissects the area of study. The gravity tool in the present study is applied on two studied areas. The first area is located at the northeastern part of Greater Cairo, and the second area is located at the northwestern part of Sinai, Egypt. The result of interpretation for gravity data for the first area indicates that there are many directions for fault elements of NE-SW parallel to Gulf of Aqaba and NE-SE parallel to Gulf of Suez and, also, the E-W trend parallel to the trend of Mediterranean Sea. The result of interpretation of the second area indicates different fault elements of different direction such as NW-SE trend parallel to trend Gulf of Suez, NE-SW is parallel to Gulf of Aqaba and E-W trend parallel to the trend of Mediterranean Sea.

Keywords: fault, structural trends, Cairo, Sinai, Egypt

1. Introduction

Geophysical tools are applied to delineate geological structures, to investigate promising areas, and to locate ore bodies by clarifying the distributions of physical properties in the earth. The main targets for gravity survey are mainly used for delineating subsurface structures which control the configuration of oil reservoirs and groundwater aquifers. Gravity survey can be used for geotechnical investigation through determining the subsurface features such as caves, fractures and faults. Gravity methods also can be used for mineral exploration through the interpretation of gravity data that depend on the density contrast between the ore bodies and contrary rock. The gravity data interpretation can used for archeological prospecting to locate the archeological features which are buried at different historical times. Many authors used the gravity data interpretation for groundwater exploration such as [1, 2]. Araffa and Fernando [3] used gravity data to delineate structures and tectonic on northern part of Greater Cairo.

2. Gravity data acquisition

The gravity data collected in two areas by using CG3 Instrument of spacing between stations of 3–5 km according to ease of roads, wadis and tracks. The CG3 Autograv is a microprocessor-based automated gravity meter that has a measurement range of over 7000 mGals, a reading resolution of 0.005 mGals.

3. Gravity data corrections

The measured gravity data are reduced for different corrections such as tide, drift, latitude, free air, and Bouguer anomaly. Also, gravity data are corrected for topographic correction where the two study areas are rough topography and then have direct effect on the gravity measurements using Hummer template.

4. Case studies

The gravity measurements were carried out on different localities in Egypt for delineating subsurface structures which have been controlled the configuration of groundwater aquifers.

4.1. Area Northeastern part of Greater Cairo

The Greater Cairo consists of three governorates: Cairo, Giza, and Kalubia. The Greater Cairo is located in northern Egypt, known as Lower Egypt. Although the Cairo metropolis extends away from the Nile in all directions, the governorate of Cairo resides only on the eastern bank of the river in addition to two islands within it on a total area of 214 km^2 (**Figure 1**). Greater Cairo is affected by different fault elements [4–6]. Araffa [7] evaluates the structural elements in Mokattam area (east Cairo). The present study aims to detect the fault elements and their role on the distribution of the basaltic sheet, to determine the basins and uplifts, as well as to estimate groundwater and oil potentiality. The last target of this study is to evaluate the active fault elements and potential effects on the environment through population and constructions.

4.1.1. Geology of the study area1

The geological setting of the study area is investigated through the geological map which is published by [8] (**Figure 1b**). The geological map shows that the cultivated parts are occupied by the western part of the study area. Geologically, the study area contains different geological units of different ages such as Quaternary, Middle Miocene, and Oligocene deposits. Different formations that belong to Quaternary deposits are represented in the study area and mainly composed of sand sheets which are located at the eastern part of the area and sand dunes which are found at the central and southern part of the area.

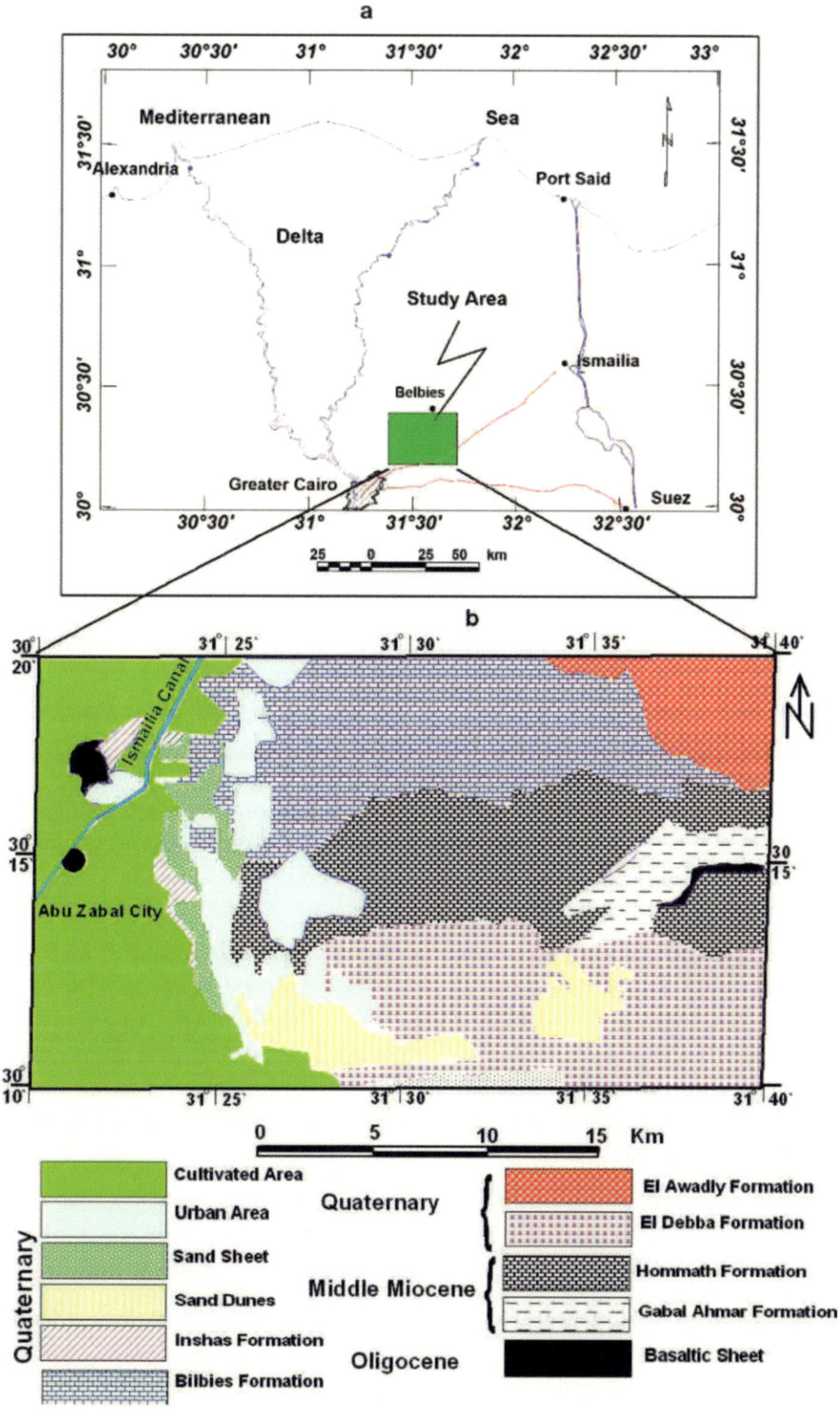

Figure 1. (a) Location map of the area1 and (b) geologic map of the study area1 (Araffa et al, 2014).

While Inshas Formation is composed of sand, intercalated with Nile mud and silt, occupies the eastern part of the study area, the next formation of Quaternary deposits is the Bilbies Formation which consists of sand and carbonate pockets located at the northern part of the area. El Awadly Formation is located at the northeastern part and composed of cross bedded gritty sands. The last formation of Quaternary deposits in the study area is represented by El Debba Formation which covers the southeastern part of the area and is composed of sand intercalated with flint. The Middle Miocene deposits are represented by Hommath Formation, which are composed of sandy limestone, sandstone, and sandy marl, occupying the central part of the survey area. The Oligocene deposits are represented by two formations: the first one is Gabal Ahmar Formation located at the eastern part of the Greater Cairo and consists of fine sand and sandstone, while the second formation is represented by basaltic sheet which is located at the western part of Greater Cairo and lies at the north of Abu Zabal city.

4.1.2. Gravity data and interpretation

Two thousand two hundred and fifty stations are carried out to cover the study area using CG-3 Autograv (**Figure 2a**). The gravity anomalies observed in the Bouguer field are caused by lateral density contrasts within the sedimentary section, crust, and subcrust of the earth. The corrected gravity data is then used to construct a Bouguer anomaly map (**Figure 2b**) using [9].

4.1.3. Regional/residual separation of gravity data at area1

The regional-residual separation technique is carried out to filter the regional component, which originates due to deep-seated sources from the residual component, which is related to local, shallow structures. Least squares technique is used to perform the gravity field's separation [10]. Calculations for different orders up to the fifth order were used to estimate the best order of separation. The results of the regional-residual separation for different orders were mapped by [9]. **Figure 3a–e** represents the regional component for the first, second, third, fourth, and fifth orders, respectively.

Figure 4a–e represents the residual component for the first, second, third, fourth, and fifth orders, respectively. A correlation factor r(x,y) has been calculated to select the best order of residual for gravity interpretation. **Table 1** shows that the best order for gravity separation is the third order. The residual gravity map of the third order was used for gravity interpretation to delineate the structural elements that dissect study area. The fault map (**Figure 5**) reveals different structural trends such as NW-SE, NE-SW, and the E-W trends parallel to the Gulf of Suez, Gulf of Aqaba, and Mediterranean Sea, respectively.

4.1.4. Conclusion of area1

From the quantitative interpretation of the gravity data, we can conclude that:

The fault elements dissecting the area are active and have characteristic trends of NW-SE, NE-SW, and E-W.

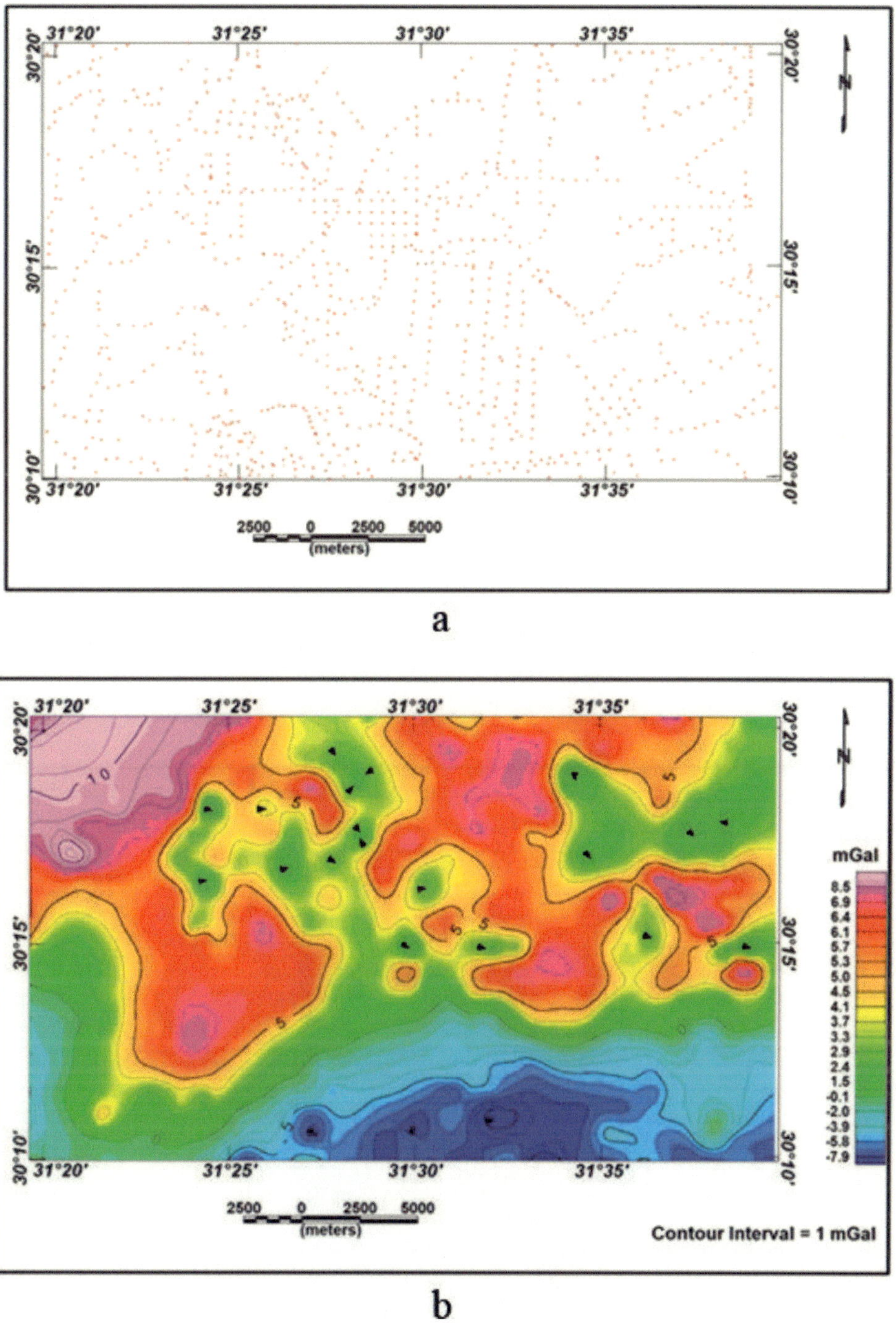

Figure 2. (a) Location map of measured gravity stations and (b) Bouguer anomaly map at area1.

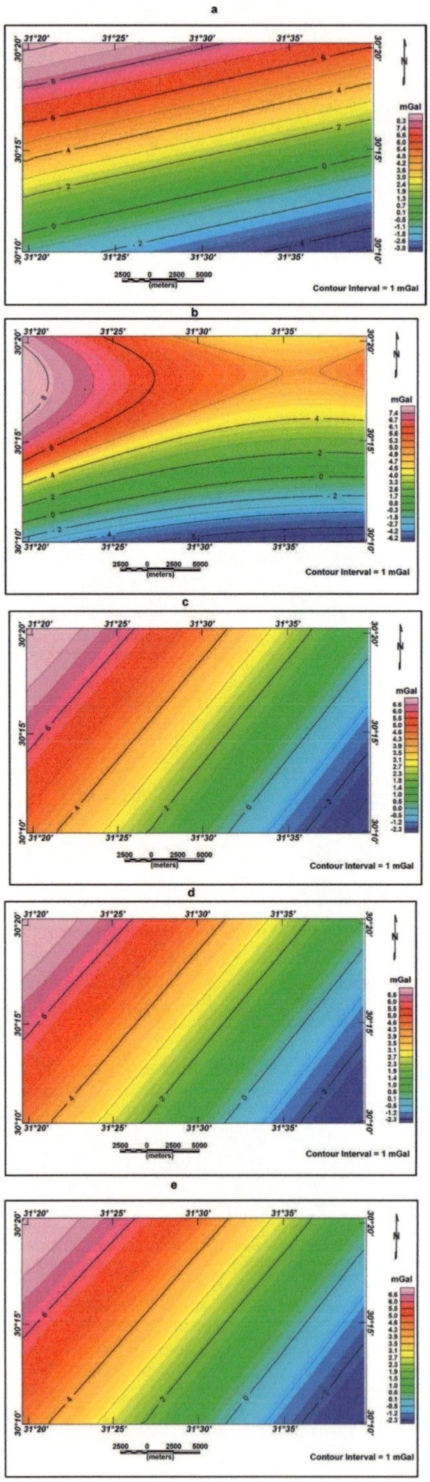

Figure 3. Regional maps for the (a) first order, (b) second order, (c) third order, (d) fourth order, and (e) fifth order.

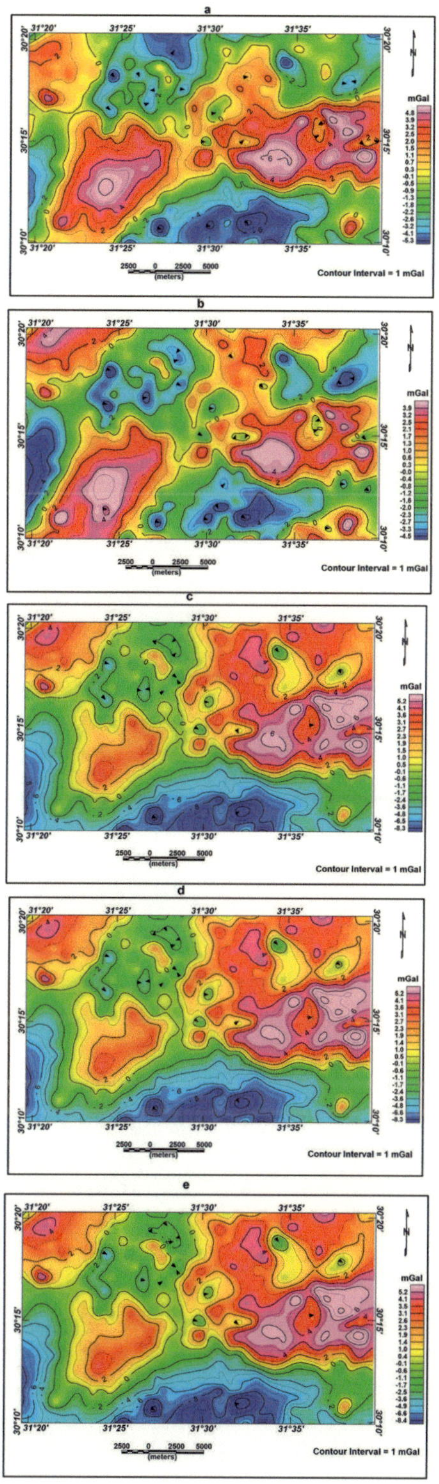

Figure 4. Residual maps for (a) first order, (b) second order, (c) third order, (d) fourth order, and (e) fifth order at area1.

Order	Value
r12	0.919825
r23	0.666593
r34	0.999966
r45	0.999966

Table 1. Correlation factor for different orders.

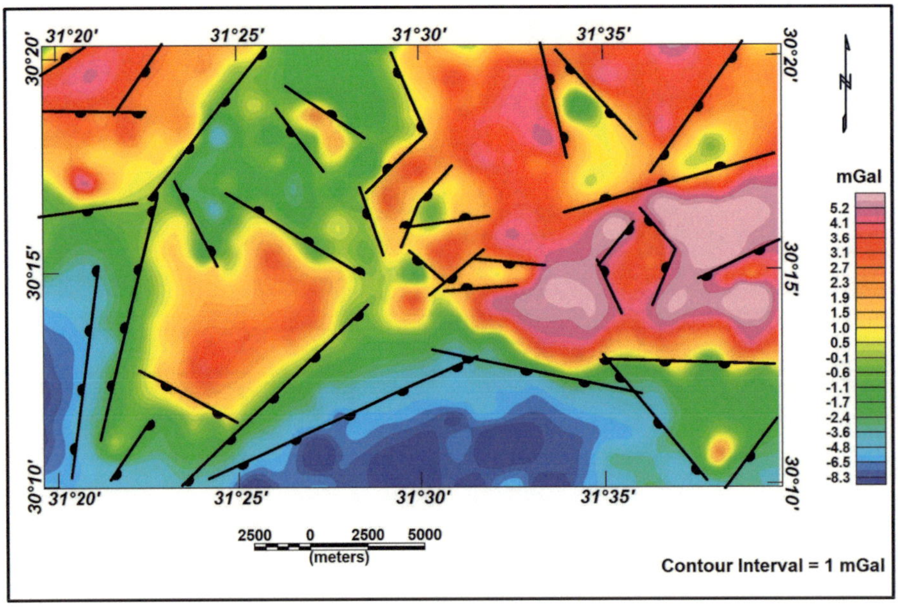

Figure 5. Location of possible faults as inferred from the gravity map of the study area1.

4.2. Area2: Northwestern Sinai, Egypt

Area2 is located at the eastern bank of the Suez Canal of northwestern part of Sinai around Al Qantara East and lies at latitudes 30° 29′ 35″ and 30° 56′ 43″ N and longitudes 32° 22′ 34″ and 32° 43′ 06″ E and represents an area of 1648 km^2 (**Figure 6**).

4.2.1. Geology of the area2

The surface geology of area2 is a part of geological map of Sinai which is constructed by [11]; the study area contains different geological units of different geological ages such as deposits of Quaternary age which include two sub-ages such as Holocene and Pleistocene. The deposits of Pleistocene contain two formations; the first one is Sabakha, which outcrops at the western part of the area2 beside the bank of Suez Canal and at the northwestern part of the area2. The second formation is the sand sheet and sand dunes which cover most the study area2. The deposits of Pleistocene in the study area include two formations (Al Qantara and

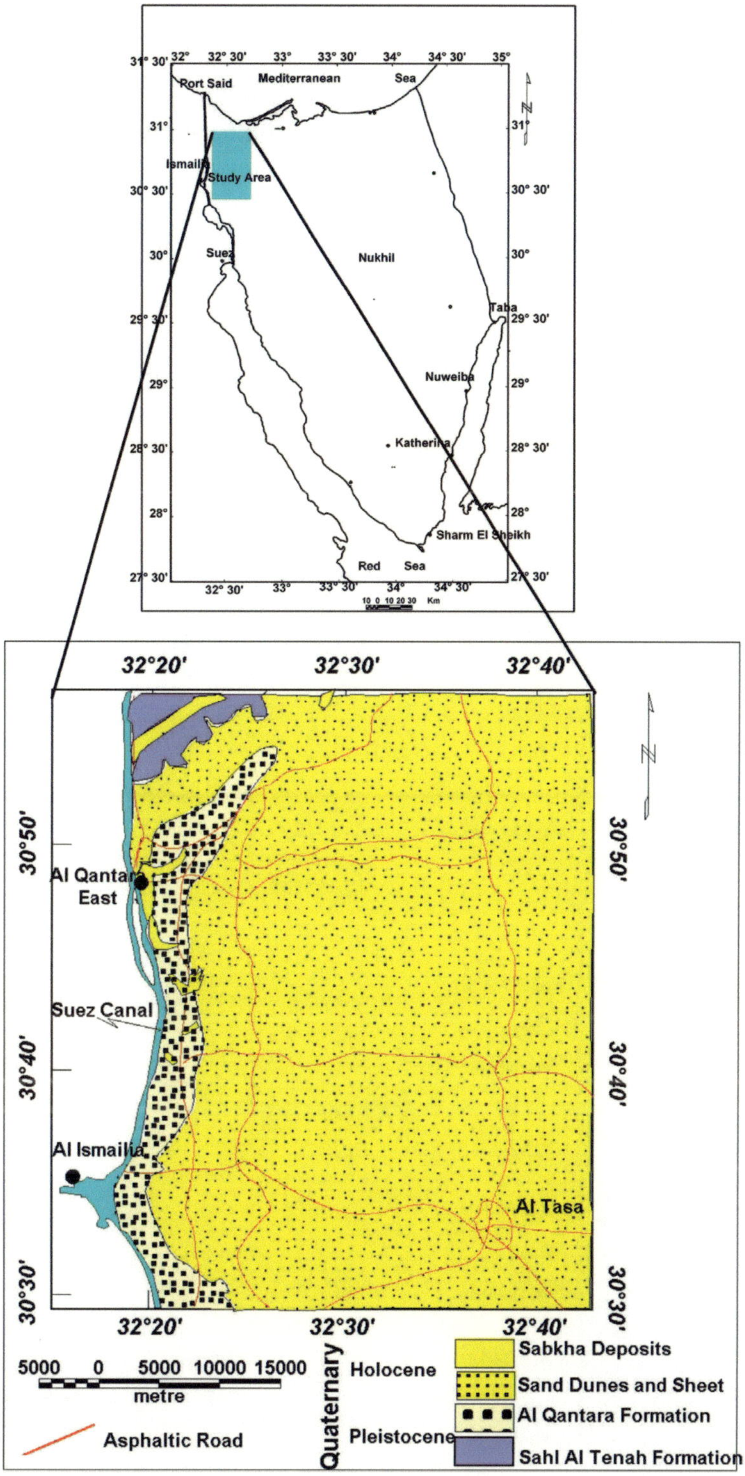

Figure 6. Location of geologic map of area2 (modified after UNESCO Cairo Office [11]).

Sehl Al Tinah). Al Qantara Formation consists silty sand with clay intercalation. Sehl Al Tinah Formation is composed of sand and silt (**Figure 6**).

The subsurface stratigraphy of the study area2 was denoted from the borehole drilled at distance of 60 km south the study area and located by latitude 29° 50′ 31″ and longitude 32° 39′ 36″ of depth 500 m. The description of lithostratigraphy for the borehole is shown as the following, Alluvium of Quaternary deposits up to depth of 30, Middle Miocene of depth from three formations 0 to 80 m (Ayun Musa Formation), which consists of sand intercalated with clay. The Lower-Middle Miocene deposits include two formations (Abu Rudeis and Nukhul). The Abu Rudeis Formation is located at depth from 80 to 230 m and represented by mud and clay. The Nukhul Formation is formed by clay and sandstone. The Lower Cretaceous is represented by Malaha Formation which consists of sandstone intercalated with clay. The Malaha Formation is represented by deep Nubian sandstone aquifer in the area2 (**Figure 7**).

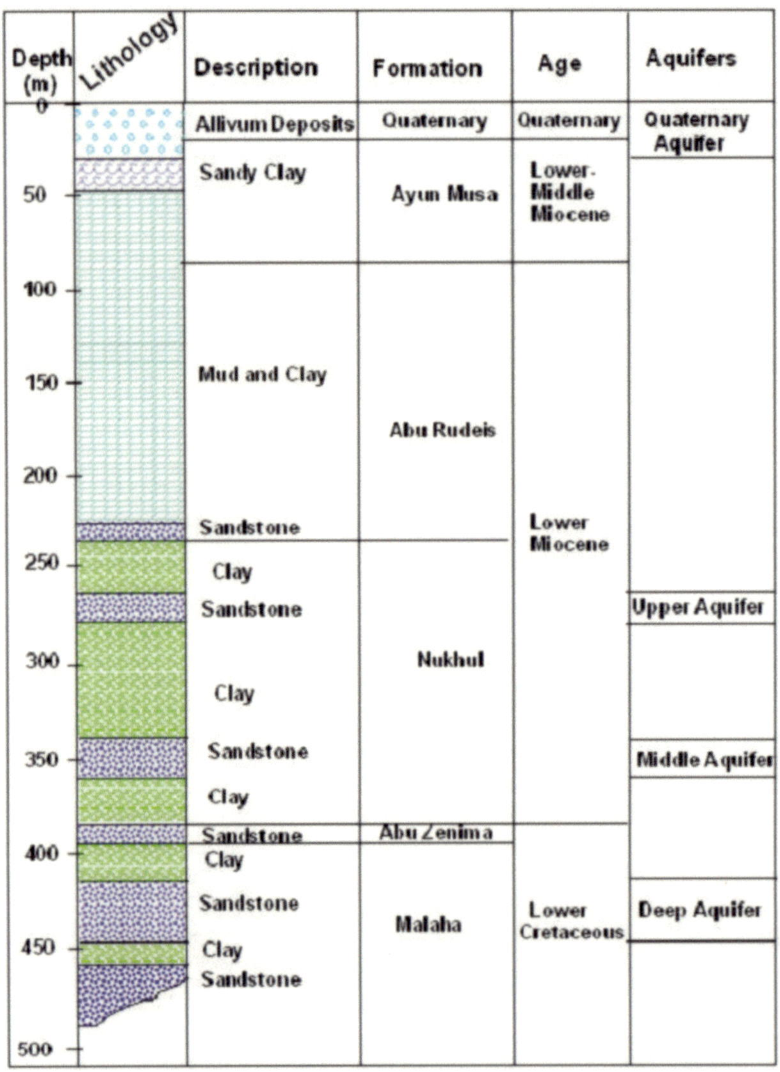

Figure 7. Al Qantarah borehole drilled at Al Qantarah Coast Village.

4.2.2. Gravity measurements and interpretation of area2

Sixty-five gravity stations have been done using gravimeter of model CG3 Autograv. The collected gravity data were corrected to different gravity corrections such as drift, tide, free air, Bouguer, latitude, and topographic corrections using specialized software [9]. The corrected gravity values were resented by Bouguer anomaly map using [9]. The gravity anomaly map (**Figure 8**) reflects different high and low anomalies according to density variations of subsurface rocks. The gravity anomaly map separated into two components (regional and residual which are related to deep and shallow sources). The least squares technique was applied for gravity separation. **Table 2** indicates that the calculations of least squares technique are carried out up to the fifth order and reflect that the best order for gravity separation is the fourth order. **Figures 9–13** represented the regional and residual for the orders from the first to fifth. The fourth order residual gravity map

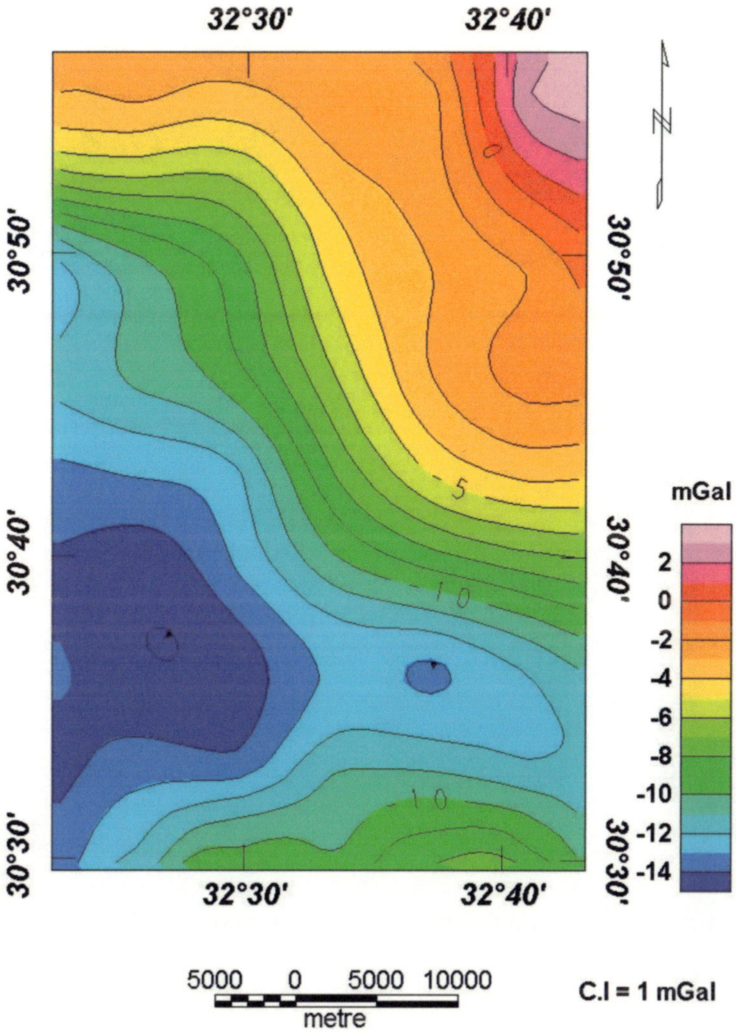

Figure 8. Bouguer anomaly map of area2.

Order	Value
r12	0.7584
r23	0.4728
r34	0.25965
R45	0.9999

Table 2. Correlation factor for different orders.

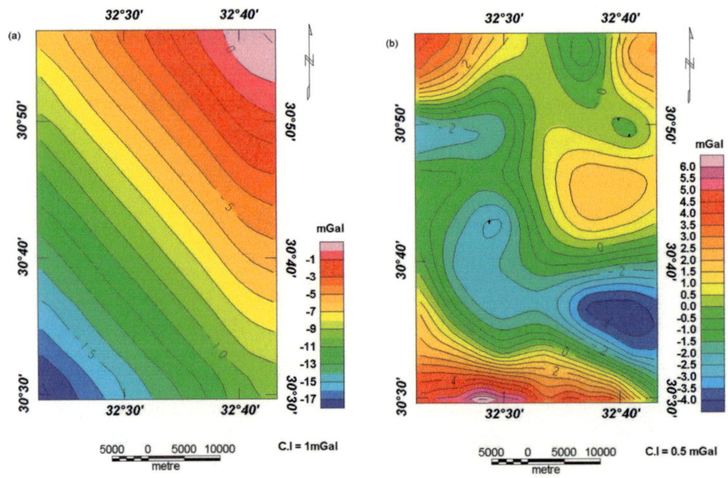

Figure 9. (a) First-order regional Bouguer anomaly map and (b) first-order residual Bouguer anomaly map.

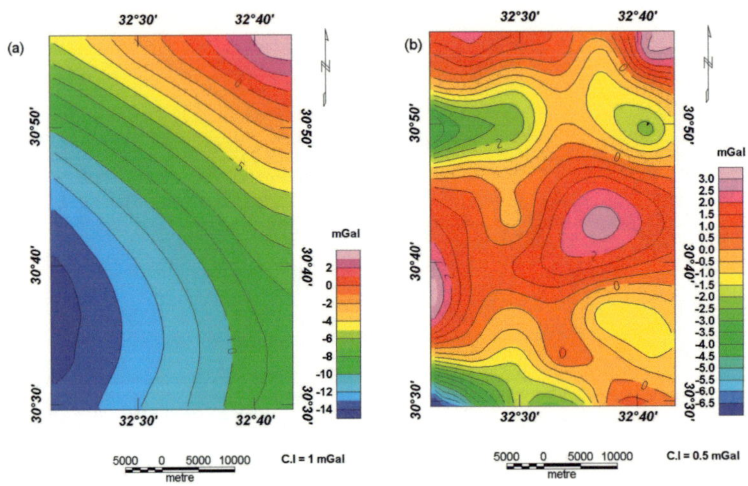

Figure 10. (a) Second-order regional Bouguer anomaly map and (b) second-order residual Bouguer anomaly map.

http://dx.doi.org/10.5772/intechopen.71492

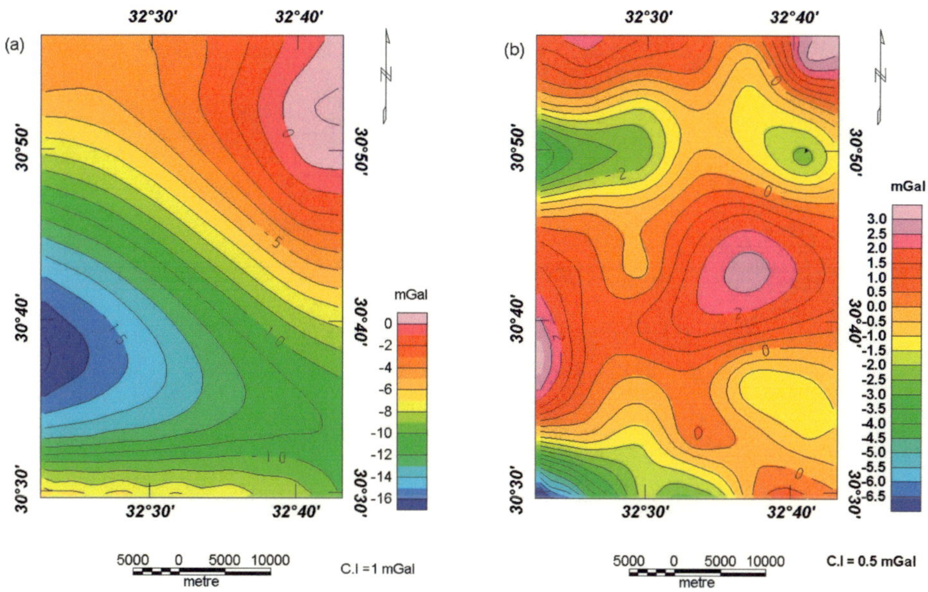

Figure 11. (a) Third-order regional Bouguer anomaly map and (b) third-order residual Bouguer anomaly map.

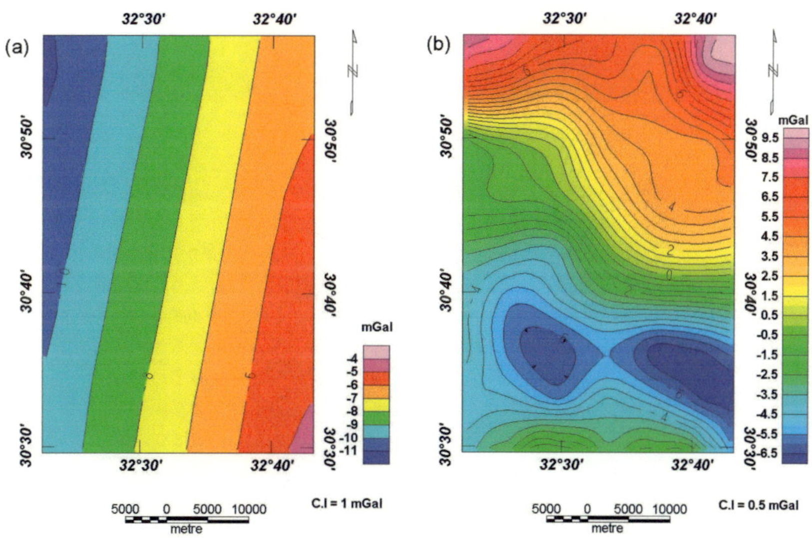

Figure 12. (a) Fourth-order regional Bouguer anomaly map and (b) fourth-order residual Bouguer anomaly map.

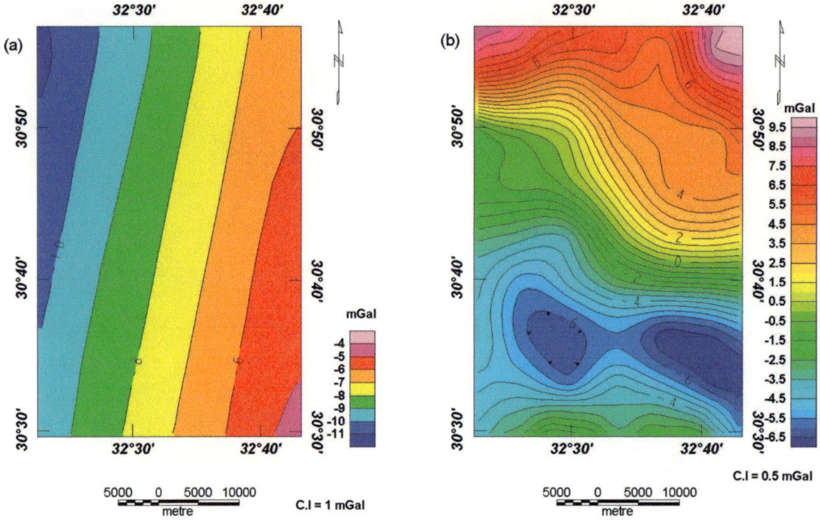

Figure 13. (a) Fifth-order regional Bouguer anomaly map and (b) fifth-order residual Bouguer anomaly map.

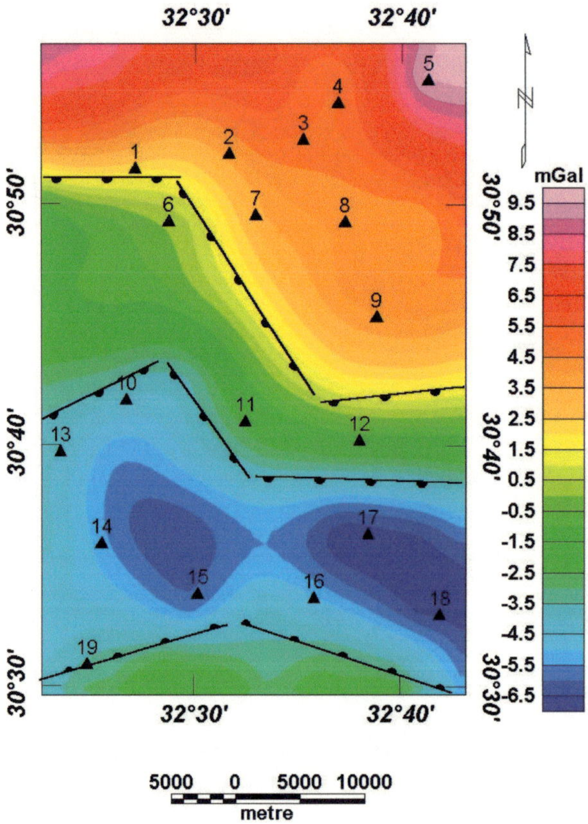

Figure 14. Fault elements dissecting study area from gravity interpretation.

was used for gravity interpretation to delineate the fault elements that dissect study area and controlled the geometry of the groundwater aquifers. **Figure 14** represents the fault map in the study area which shows different fault elements of different directions such as NW-SE trend parallel to trend Gulf of Suez; NE-SW is parallel to Gulf of Aqaba and E-W trend parallel to the trend of Mediterranean Sea. The result of gravity interpretation indicated that the study area is dissected by different fault elements of trends NE-SW, NW-SE, and E-W.

Author details

Sultan Awad Sultan Araffa

Address all correspondence to: sultan_awad@yahoo.com

National Research Institute of Astronomy and Geophysics, Helwan, Cairo, Egypt

References

[1] Araffa SAS, Sabet HS, Gaweish WR. Integrated geophysical interpretation for delineating the structural elements and groundwater aquifers at central part of Sinai Peninsula, Egypt. Journal of African Earth Sciences. 2015;**105**:93-106

[2] Sultan SA, Mohameden MI, Santos FM. Hydrogeophysical study of the El Qaa Plain, Sinai, Egypt. Bulletin of Environmental Geology and Engineering. 2009;**68**:525-537

[3] Araffa SAS, Santos FAM, Arafa-Hamed T. Assessment of subsurface structural elements around greater Cairo by using integrated geophysical tools. Environment and Earth Science. 2014;**71**:3293-3305

[4] Sultan SA, Santos FAM, Helaly AS. Integrated geophysical analysis for the area located at the eastern part of Ismailia Canal, Egypt. Arabian Geosciences Journal. 2011;**4**:735-753

[5] Sultan SA, Santos FAM. Evaluate subsurface structures and stratigraphic units using 2-D electrical and magnetic data at the area north Greater Cairo, Egypt. International Journal of Applied Earth Observation and Geoinformation. 2008;**10**:56-67

[6] Santos FM, Sultan SA. On the 3-D inversion of vertical electrical soundings: Application to the South Ismailia – Cairo Desert Road area, Cairo, Egypt. Journal of Applied Geophysics. 2008;**65**:97-110

[7] Araffa SAS. Geophysical investigation for shallow subsurface geotechnical problems of Mokattam Area, Cairo, Egypt. Environmental Earth sciences journal. 2010;**59**: 1195-2207

[8] Geological Survey of Egypt (EGSMA). Geology of Inshas Area, Geological Survey of Egypt. Internal report: 1998

[9] Montaj O. Geosoft mapping and application system, Inc, suit 500, Richmond St. West Toronto. In: ON Canada N5SIV6. 2007

[10] Abdelrahman EM, Refai E, Amin Y. On least-squares residual anomalies determination. Geophysics. 1985;**50**(3):473-480

[11] UNSECO Cairo Office. Geologic Map of Sinai, Egypt, Scale 1:500,000, Project for the Capacity Building of the Egyptian Geological Survey and Mining Authority and the National Authority for Remote Sensing and Space Science in Cooperation with UNDP and UNSECO. Geological Survey of Egypt. 2004

Section 3

Gravity in Industrial Technology

Chapter 7

Gravity in Heat Pipe Technology

Patrik Nemec

Additional information is available at the end of the chapter

http://dx.doi.org/10.5772/intechopen.71543

Abstract

This work deal with heat pipes experiments depending on the gravity, because gravity is one of the phenomenon which affects the heat transport ability of the heat pipe. In heat pipe technology is gravity on the one side positive when ensure the circulation of the working fluid in gravity heat pipe (GHP) (thermosiphon) and on the other side negative when act opposite the capillary action in the wick heat pipe (WHP). The work describes principle, design and construction of gravity heat pipe (GHP) and wick heat pipe (WHP). Experimental determination influences gravity on the heat transport ability of gravity heat pipes by changing working position. Experimental and mathematical determination influences gravity on the heat transport ability of wick heat pipes by changing working position.

Keywords: heat pipe, wick structure, gravity, capillary action, heat transfer, thermal performance, mathematical model

1. Introduction

Heat pipe technology is one form of technology which is mostly used in technical application. It is mainly used in area of heat transfer such as electronic cooling, heat recovery from technical processes, recovery of heat from ground and sun for use in the heating system.

The principle of two-phase heat transfer with liquid phase transport due gravity first use Angier March Perkins in device known as a Perkins Tube in the middle nineteenth century. Today is Perkins invention known as a Thermosiphon. The capillary effect used in liquid phase transport in two-phase heat transfer device was first suggested by Gaugler in 1942 under the name Heat Pipe [1].

The basic heat pipe or thermosiphon consists of hermetically sealed pipe in which the working medium is at a defined pressure. Heating one end (evaporation part) of the pipe and cooling the other end (condensation part) cause the working medium to evaporate. Vapour flow through the adiabatic part of heat pipe into the condensation part which condenses in to liquid and returns back to the evaporation part via gravitational force or other forces (e.g. capillary action, centripetal force, osmosis or electrohydrodinamics) which evaporates again. This creates the natural circulation of the working medium associated with heat transfer. In the case of heat transfer with a phase change of the working medium, the latent heat of the substance is released and hence the heat transfer efficiency through the heat pipe is very high [2].

Due to very high heat transfer coefficients for boiling and condensation, these devices are highly effective thermal conductors over many other heat-dissipation mechanisms. They contain no mechanical moving parts and typically require no maintenance. The other advantages of devices with heat pipe technology are a fast start action of the heat transfer, their geometric shape and weight [3]. In **Figure 1**, schema of heat pipe is shown.

Both thermosiphon and heat pipe are heat transfer devices which work on the same principle. The difference is only in the method of the liquid phase transport. In one device, the gravity acts positive and in other acts negative [4].

The gravitational force is an attractive force, which acts between two masses, two bodies or two particles. Gravity not acts only between objects on the Earth, but also acts between all objects, anywhere in the Universe. Sir Isaac Newton found that to change the speed or direction of movement of the object, gravity force is required. He also found that gravity force causes the apple to fall from the tree, or humans and animals live on the surface of our planet rotating around its own axis without being thrown away. He further deduced that there are gravitational forces among all objects. Newton's gravity law is a mathematical description of how the bodies attract each other, based on experiments and observations performed by scientists. The mathematical expression of the Newton's law is

$$F = \frac{Gm_1m_2}{r^2} \tag{1}$$

where G is the gravity constant and it value is $6.6726 \times 10^{-11}\ m^3\ kg^{-1}\ s^{-2}$.

It claims that the gravity is proportional to the product of two masses (m1 and m2) and is inversely proportional to the square of the distance (r) between its centres of gravity. The

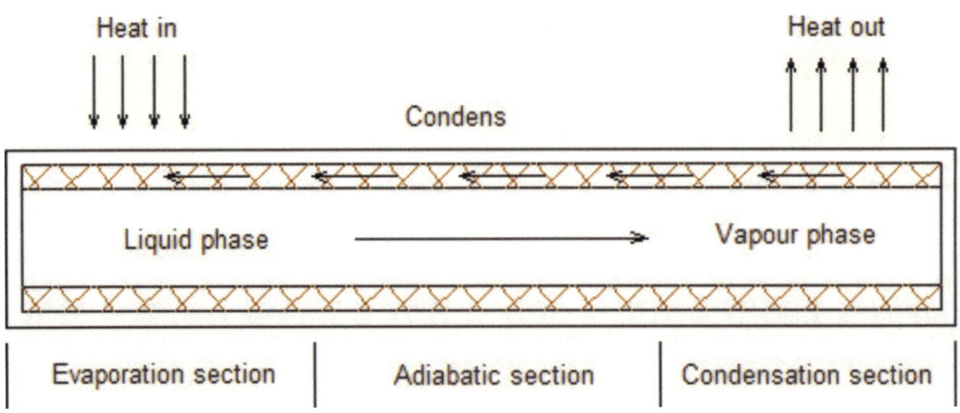

Figure 1. Principle of heat pipe.

influence of the gravitational force goes from every object into the space in all directions and at an infinite distance. However, the gravitational force rapidly decreases with a distance.

Capillary action (or capillarity) describes the ability of a liquid to flow against gravity in a narrow space such as a thin tube.

This spontaneous rising of a liquid is the outcome of two opposing forces:

Cohesion—the attractive forces between similar molecules or atoms, in our case, it is between the molecules or atoms of the liquid. Water, for example, is characterised by high cohesion since each water molecule can form four hydrogen bonds with neighbouring molecules.

Adhesion: the attractive forces between dissimilar molecules or atoms, in our case the contact area between the particles of the liquid and the particles forming the tube.

In view of molecular physics, capillarity is caused by capillary pressure. Consider the liquid with surface tension that perfectly wet the walls of the vessel. When the capillary with the internal radius is immersed into the liquid, a concave hemispherical surface of radius is formed in the capillary. The internal pressure under concave surface in capillary is less about the capillary pressure compared to the horizontal surface in the larger vessel. This results in the liquid rising in the capillary to such a height that the hydrostatic pressure corresponding to the height of the liquid column is the same as the capillary pressure. The height *(h)* of a liquid inside a capillary is given by the expression

$$h = \frac{2\gamma cos\theta}{\rho g r} \tag{2}$$

where γ is the liquid air surface tension, θ is the contact angle, ϱ is the density of the liquid, g is the gravitational field and r is the tube radius.

In case, if the forces of adhesion are greater than those of cohesion and gravity (when it exists), the molecules of the liquid cling to the wall of the tube. We will observe that the upper surface of the liquid becomes concave (the height of the liquid at the contact area is higher than its height at the centre of the tube). The cohesive forces between the molecules of the liquid are "attempting" to reduce the surface tension (i.e. to flatten the upper surface of the liquid and thus prevent the increased surface area in the concave state). In doing so, the molecules keep climbing up until a steady state between cohesion and adhesion is achieved (with or without the gravity component). This also explains why this phenomenon occurs exclusively in thin tubes (also in the absence of gravity). In wider vessels, only a small fraction of the liquid comes into contact with the vessel walls, and so adhesive forces are negligible and there is hardly any rising of the liquid.

2. Heat pipe construction

The heat pipe may have several basic parts depending on its type. During the heat pipe development, the main components and materials remained the same. The simplest type of heat pipe consists of two basic parts, the body (container) and the working medium. A capillary structure (wick) can be placed inside the heat pipe body to allow the condensed liquid phase of the working fluid wicking against the vapour flow due the capillary action. Such a heat pipe is called a wick heat pipe (WHP). The heat pipe without capillary structure is called gravitational heat pipe because it returns the liquid phase from the condenser part to the evaporator part which is due to gravity [5].

2.1. Heat pipe container

The container of the heat pipe can have for different applications different shapes, but the most often is in the form of a closed pipe of a circular, flat or triangular cross-section. The main function of the heat pipe container is to isolate the working fluid from the outside environment. The container of the heat pipe should be strong enough to prevent internal dimension and internal pressure in case of compression or bending. The choice of the container material depends on many of its properties and should have the most appropriate combination (working fluid and environment compatibility, strength-to-weight ratio, thermal conductivity, porosity, wettability, machinability, formability, weldability or bondability).The container material should have a high thermal conductivity, solid and tough but easily machined, formable and easily soldered and welded. The surface of the material should be well wetted, but at least as porous as possible to avoid gas diffusion. The materials most commonly made of heat pipes are steel, copper, aluminium and their alloys. Various coatings of steel materials are also used [6].

2.2. Working fluid

Since the operation of the heat pipe is based on evaporation and condensation of the working fluid, its selection is an important factor in the design and manufacture of the heat pipe. The working fluid is chosen in particular according to the working temperature range of the heat pipe. Therefore, when selecting a working fluid, it is necessary to be careful if the operating temperature range of the working fluid lies in the operating temperature range of the heat pipe. The heat pipe can operate at any temperature that is in the range between the triple and the critical point of the working fluid. In case of several working fluids having the same working temperature is the decision criterion appropriate combination of its thermodynamics properties. The recommended features that working fluid should have are compatibility with the capillary structure material and the heat pipe container, good thermal stability, wettability of the capillary structure and heat pipe container, vapour pressure in the operating temperature range, high surface tension, low viscosity of the liquid and vapour phase, high thermal conductivity, high latent heat of vaporisation, acceptable melting point and solidification point [6].

2.2.1. Heat pipe classification by working fluid

The heat pipe is able operate at any temperature between triple state and critical point of the working fluid, but in the area near this point, heat pipe reduced its ability to transfer heat due to characteristic properties of viscosity and surface tension of the liquid phase working fluid. In **Table 1**, typical heat pipes working fluids are sorted by operating temperature range. The application of heat pipe depends mainly on operating temperature range of the working fluid. Depending on it, heat pipe is divided into four groups:

http://dx.doi.org/10.5772/intechopen.71543

1. *Cryogenic heat pipe* has an operating temperature range of 1–200 K. The gases are used as a working fluid in cryogenic heat pipes, for example, helium, argon, oxygen, neon or nitrogen.
2. *Low-temperature heat pipe* has an operating temperature range of 200–550 K. Working substances such as methanol, ethanol, ammonia, acetone or water are used.
3. *Medium temperature heat pipe* has an operating temperature range of 550–700 K. Working fluids such as mercury, sulphur or some organic liquids (naphthalene, biphenyl) are used.
4. *High-temperature heat pipe* has an operating temperature range in the temperature range above 700 K. As a working fluids are used metals such as potassium, sodium, silver, which melt at high temperatures and are in the liquid state [5].

In practice, water is the most often used working fluid in heat pipe, because it has optimum operating temperature range, is the cheapest, available, has the best thermodynamics properties and the highest latent heat of vaporisation, which is its great advantage in heat transfer compared with the other fluids. However, due to the high pour point, it is not suitable for low-temperature operations outside and is not compatible with all materials. In low-temperature operations, the often used fluids are ethanol, ammonia and CO_2. In high-temperature operations, suitable liquid metals with a low melting point such as sodium and potassium are used. Some fluids have been stopped to use due to increased health, safety and environmental requirements. For example, the use of CFCs in European countries is forbidden, and the use of HFC halogenated hydrocarbons is gradually being phased out in favour of fluids that have a less impact on the environment.

Working fluid	Melting point at atmospheric pressure (°C)	Boiling point at atmospheric pressure (°C)	Latent heat of vaporisation (kJ.kg^{-1})	Useful range (°C)
Helium	−271	−269	21	−271 to −269
Nitrogen	−210	−196	198	−203 to −160
Ammonia	−78	−33	1360	−60 to 100
Acetone	−95	57	518	0–120
Methanol	−98	64	1093	10–130
Ethanol	−112	78	850	0–130
Water	0	100	2260	30–200
Mercury	−39	361	298	250–650
Caesium	29	670	490	450–900
Potassium	62	774	1938	500–1000
Sodium	98	895	3913	600–1200
Lithium	179	1340	19,700	1000–1800
Silver	960	2212	2350	1800–2300

Table 1. Typical heat pipe working fluids.

2.3. Wick structures

The wick structure is placed on the inner surface of the heat pipe so that the heat pipe can operate both in the horizontal position and in the position with the evaporator located above the condenser. The wick structure function in heat pipe is to transport the condensed liquid phase of the working fluid from the condenser part to the evaporator part of the heat pipe. Another function is to ensure a uniform temperature distribution throughout the evaporator surface. There are several types of wick structures but neither is ideal. Each wick structure has some advantages and disadvantages. Therefore, when choosing a wick structure, it is necessary to consider the conditions in which the heat pipe will operate. The choice of the wick structure of the heat pipe depends on many factors. Some of them are closely related with the properties of the working fluid, working conditions and compatibility of the heat pipe components. The most important factor is ability to create capillary pressure required to working fluid transportation from the condenser section to the evaporator section where the heat pipe receives heat. The maximum capillary pressure produced by the wick structure increases with decreasing pore size. The permeability of the wick structure increases with increasing porosity. Thickness of the wick structure is another factor to take care, because the heat transfer capacity of the heat pipe depends on the wick structure thickness. Generally, the wick structure of heat pipe should have small capillary radius, high porosity, high thermal conductivity and high permeability [6].

2.3.1. Wick structure classification

Depending on whether the wick structure of a heat pipe is made from one or more structures, it is divided into homogeneous or non-homogeneous (composite). The homogeneous wick structure is created only by one structure. In **Figure 2**, typical homogeneous wick structures are shown. The mesh screen wick structure shown in **Figure 2a** consists of metal meshes located on the inner surface of the pipe. The vapour flow resistance of the working fluid depends on the tightness of the mesh wrapping in the heat pipe. The porous wick structure shown in **Figure 2b** is created from the porous material made by sintering metal powders or ceramic and carbon foams and felt. Compared to other structures, the porous wick structure has a higher effective thermal conductivity. The grooved wick structure shown in **Figure 2c** is formed by cutting the grooves into the inner surface of the heat pipe material. It creates sufficient capillary pressure, does not impose resistance to the vapour flow and preserves the thermal conductivity of the material [5].

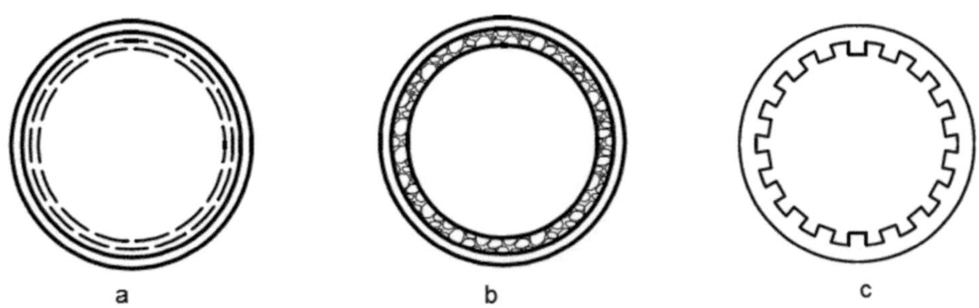

Figure 2. Cross-sections of homogenous wick structures. (a) Wrapped screen, (b) sintered metal, (c) axial groove.

http://dx.doi.org/10.5772/intechopen.71543

The composite wick structure is made of two or more kinds wick structures. In **Figure 3**, typical composite wick structures are shown. By the composite wick structure, a separate flow of the liquid phase and vapour phase working fluid in the heat pipe is achieved. For example, if the wick structure with grooves on the inner wall are covered by a metallic mesh, high capillary pressure is created and axial grooves reduces flow resistance. Thus heat transfer capacity of the grooved wick heat pipe in antigravity position can be improved.

2.3.2. *Wick structure manufacturing*

The grooved wick structure is a simple wick structure in view of production. The grooved wick structure shown in **Figure 4** is most often made by longitudinally cutting grooves in the pipe material of dimensions only a few tenth millimetre. Another possibility of producing grooved structures is the production of groove moulds by casting or sintering metal powders which are then inserted into the tube. The copper tubes manufacturers already offered a wide range of copper pipes with longitudinal grooves suitable for using in heat pipe production. In applications of this wick heat pipe, it should take in account that cutting the grooves may weaken the pipe container. Since low capillary pressure is formed in the wick structure with axial grooves, it is suitable to use them in applications where the heat pipe operates in a horizontal or in a position with an evaporator position under the condenser.

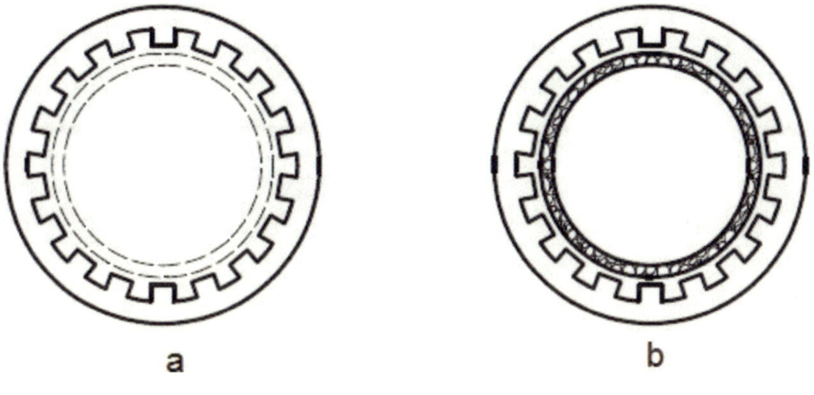

Figure 3. Cross sections of composite wick structures. (a) Axial groove – wrapped screen, (b) Axial groove – sintered metal.

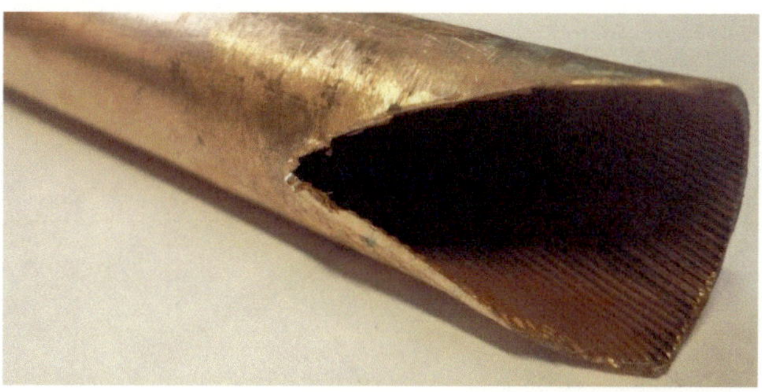

Figure 4. Grooved wick structure.

The mesh screen wick structures are among the most commonly used wick structures of wick heat pipes. It is mainly for their simply production, easy availability, wide range of materials and variability to create different combinations (number of layers and number of mesh types) according to heat flux transfer and heat pipe orientation. In general, fine-meshed anti-corrosion meshes with a mesh size range of 50–250 are used to mesh screen wick structure production. Mesh size is given by the number of holes in the screen per unit length (inch = 25.4 mm). The mesh screens are coiled into several layers so that they are firmly attached to the inner surface after insertion into the pipe by their own expansive forces. In **Figure 5**, samples of fine mesh screens from which the wick structures for heat pipes have been made are shown. In **Figure 6**, samples of mesh screen wick structures inserted in the pipes are given.

Although the production of the porous wick structure is the most difficult from all types of wick structures, is one from the three most used wick structures in the heat pipe, because is able creates a large capillary pressure that allows the heat pipe to transfer a high heat flux in the antigravity position. One method of making a porous wick structure is to sinter a copper powder uniformly poured around a coaxially centred steel mandrel located inside the copper pipe at a temperature close to melting the powder material in a high-temperature electric furnace. The formation of a suitable porous structure by sintering the metallic powder depends, in addition to the sintering temperature, both from the time of sintering and the grain size of the powder. Copper powders

Figure 5. Fine mesh screens (mesh 50, 100, 200).

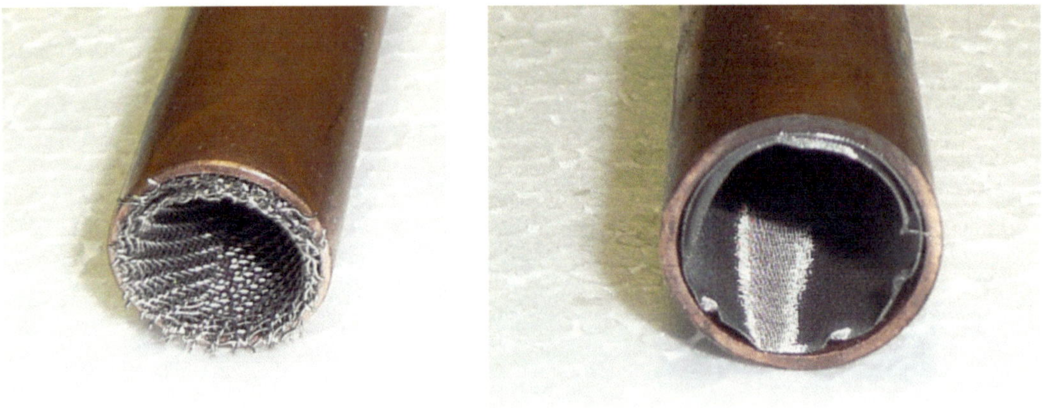

Figure 6. Mesh screen wick structures.

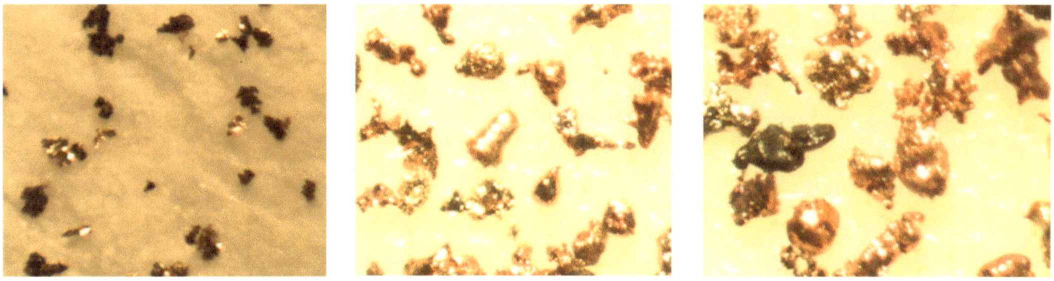

Figure 7. Granules of copper powders (35, 63, 100 μm).

Figure 8. Sintered porous wick structure.

with a particle size of 30–100 μm or copper fibbers of 2–3 mm in length and a diameter of 20 – 100 μm are used to production porous sintering structure. In **Figure 7**, the granules of copper powders used to production sintered structure are shown. In **Figure 8**, samples of sintered wick structures made of copper powder with a grain size of 63 μm are shown.

Modern technologies allow to manufacture porous structures from metallic, ceramic or carbon materials in the form of foils or felt with different pore size ranges that are suitable for condensate transport from the condenser to the evaporator in the heat pipe. Porous structures made of ceramic materials have smaller pores, but their disadvantages are that they have little stiffness and therefore must be scaffold with a metal grid. The problem of scaffold porous structure is that while the ceramic structure itself can be chemically compatible with the working fluid, the second added material may not be compatible. Recent attention has been paid to the production of porous structures made of carbon materials. Structures of carbon materials have finer grooves on the surface, which allow to produce high capillary pressure and are chemically stable [7].

3. Heat pipe manufacturing

The main requirements on the heat pipe production are the high purity of the material of the individual parts and the working substance, as well as their mutual compatibility.

The basis of the heat pipe construction is the pipe body and the working fluid. The production of a heat pipe primarily consists in selecting a suitable material of the pipe and the working fluid. The working fluid is selected according to the temperature conditions in which the heat pipe will be used, because heat flux transferred by the heat pipe depends on the material of the pipe, the working fluid and their mutual compatibility. An important part of the wick heat pipe is the wick structure, which also has a large impact on the amount of transferred heat flux.

The main components of heat pipe are as follows:

- Pipe body (container)
- Working fluid
- Wick structure
- End caps
- Filing pipe

The heat pipe body may be of any cross-section, for example, circular or square, may include mounting flanges for ease of assembly and may be bent into various shapes. The wick structure can be formed by grooves extruded into a pipe body or fine mesh screen, porous material and artery inserted into the heat pipe body [8]. **Figure 9** shows a schema of the wick heat pipe construction.

The most common shape of the heat pipe is cylinder, because in addition of easily available product (wide assortment of material and the size of the pipe cross-sections), it provides certain advantages also in terms of strength and thermomechanical parameters. The advantage of producing a cylindrical shaped heat pipe is in the ease of handling with the cylindrical material. In practice, heat pipes with flat rectangular, triangular or other cross-sections are also used. The most common heat pipes are manufactured with an inner diameter of 8–25 mm and an internal diameter of 2–5 mm—the so-called micro-heat pipes. The production process of the heat pipe can be divided into several sub-processes involving mechanical and chemical treatment of materials.

Technological process of the heat pipe production cycle:

- Body production and end caps.
- Production of wick structure.
- Cleaning of components.
- End caps closure by impervious joints (welding, soldering).
- Mechanical verification of body strength and tightness.
- Vacuuming of inner space and filling with the working fluid.
- Sealing the filling pipe (welding, soldering).

Before heat pipe production, it is necessary to thoroughly clean all components of the heat pipe to avoid any undesirable influence, which could ultimately have an effect on the heat transfer

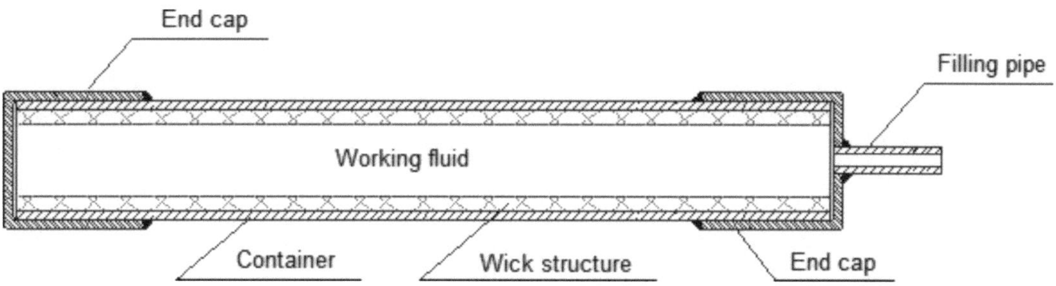

Figure 9. Schema of the wick heat pipe construction.

ability reduction. In cleaning process, first, the mechanical impurities and rust from the body of the pipe are removed manually and then chemical cleaning of the body, wick structure, end caps and filling pipe are followed [9].

3.1. Mechanical part of heat pipe production

In the mechanical part of the production, the individual components of the heat pipe are first prepared: the body, the filling pipe, the wick structure and the end caps. All components are then joined together by welding or soldering. In the case of wick heat pipe production, a wick structure is placed in the internal space of the body before heat pipe closure. The closure of the heat pipe is the connection of the body with the end caps. In **Figure 10**, the standard types of the heat pipe closure by end caps are shown. The filling pipe is connected to one of the end caps due the inner space vacuuming. After vacuuming, the heat pipe is filled with the working fluidand filling pipe is pressed, and after disconnecting from the vacuuming pump, filling pipe is sealed by soldering.

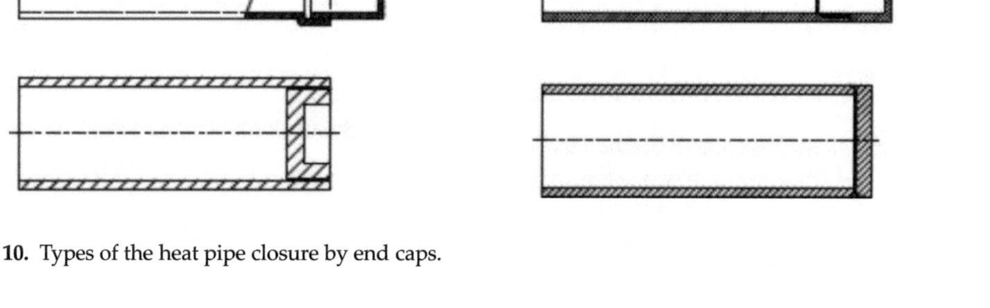

Figure 10. Types of the heat pipe closure by end caps.

3.2. Chemical part of heat pipe production

In the chemical part of the production, mechanical impurities and rust from the body of the heat pipe are first removed. This is followed by wet cleaning of the heat pipe components including cleaning with solutions, acids, basic acids which is precisely determined for each type of material. The ultrasound cleaning, vacuuming, degassing and passivation are processes that guarantee a high purity of the heat pipe material and thus contribute to long-lasting failure free operation. Generally, two important goals are achieved by cleaning. The first goal is to ensure good wetting material of the heat pipe well by working. The second goal is to remove all particles of dirt because the presence of impurities in solid, liquid or gaseous form may have an adverse effect on the heat transfer ability of heat pipe. Small particles can inhibit the formation of capillary pressure in the wick structure. Machining or human hand grease may reduce the wettability of the wick structure. Oxides formed on the walls of the wick structure may also reduce the ability of the working fluid to wet the surface. It is also highly advisable to use an ultrasonic cleaner to clean the heat pipe material, as the ultrasound breaks down, impurities are firmly absorbed on the surface of metallic particles that cannot be removed by any other way. The cleaning of the heat pipe is repeated immediately before filling with the working fluid, after connecting the body with the end caps and the filling tube. After cleaning, the tube is degassed by heating to a higher temperature and vacuuming the interior. In the case of a wick heat pipe, it is necessary to remove the oxide layers from the wick structure by chemical cleaning (e.g. solvents).

3.3. Filling the heat pipe with the working fluid

The working fluid added in to the heat pipe must be completely clean, free from all mechanical impurities and gases, as their trace residues can also react with the body material of the heat pipe and forms undesirable elements. Clean substances can be purchased without any problems at special chemical stores. However, even in pure liquids and solids, an incompressible gas may be present. These gases can be removed by repeated freezing and thawing cycles. The working fluid in the filling bottle can be frozen using liquid nitrogen or dry ice.

Filling process of each type working fluid is happening under other conditions. The characteristic of the filling process depends on the state of the working fluid at ambient temperature. If the working fluid is at the room temperature in the gaseous state (cryogenic), the filling can be carried out via a gas container of high quality. On the other hand, filling and closing of liquid-metal heat pipes is appropriate to do in the vacuum chamber [10].

The filling of low-temperature heat pipes can be carried out at room temperature without the use of any protective atmosphere. Before filling the heat pipe, it is advisable to suck the air from it to ensure removal of undesirable components contained in the materials, which could be later shown as non-condensing components. In addition, due to under pressure is the working fluid naturally enter to the heat pipe and thus the equilibrium state of the pure vapour and liquid phases at a lower pressure than atmospheric will achieve [11]. The scheme vacuuming and filling of heat pipe with the low-temperature working fluid is shown in **Figure 11**.

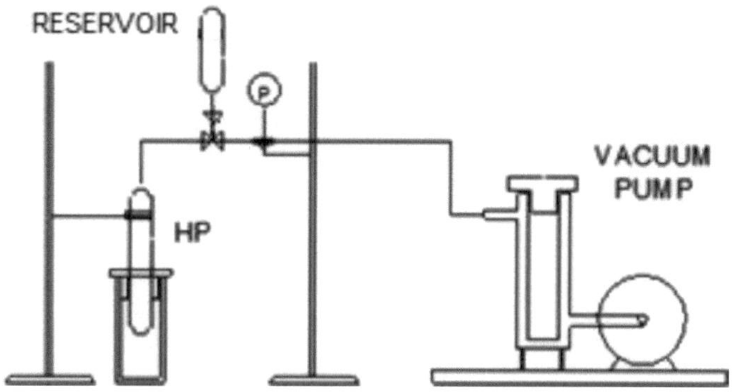

Figure 11. Schema vacuuming and filling of heat pipe.

4. Heat pipe experiments

The gravity is on the one side positive when ensure the circulation of the working fluid in gravity heat pipe (thermosiphon) and on the other side negative when act opposite the capillary action in the wick heat pipe. The gravity action in the Earth gravity field acts in one direction. If the working position of heat pipe in gravity field changed, the gravity action on the condensate flow of the working fluid also changed. Thus at the vertical position of heat pipe, gravity action can be positive or negative depending on the heat source position and at the horizontal position is the gravity action zero. Because heat transport ability of the heat pipe depends on condensate mass flow of the working fluid, the gravity is one of the main parameters which influence the heat transport ability of heat pipe. Therefore the main goal of the experiments is the determination influence of the working position (gravity) on the thermal performance amount transferred by heat pipes.

To determination thermal performance amount transferred by heat pipe was proposed measuring device consisted from measuring apparatus (rotary board, heater with heating thermostat, cooler with cooling thermostat, flowmeter, temperature sensors and data logger) shown in **Figure 12**. Thermal performance of heat pipe measurement consists of heating the evaporation section by a circulating medium at 80°C and cooling the condensation section by a circulating medium at 20°C. Rotating the rotary board is possible set up the working position of heat pipe. To determine thermal performance of heat pipe was used calorimetric method emanating from calorimetric equation where known mass flow, specific heat capacity, input and output temperature of circulating cooling medium.

$$Q = \dot{m}\,.c.\Delta t \tag{3}$$

$$\Delta t = t_2 - t_1 \tag{4}$$

where is Δt (°C) is the temperature gradient, t_1 (°C) is the input temperature, t_2 (°C) is the output temperature, $\dot{m}$ ($J.kg^{-1}.K^{-1}$) is the mass flow of liquid and c ($J.kg.s^{-1}$) is the special thermal capacities of liquid.

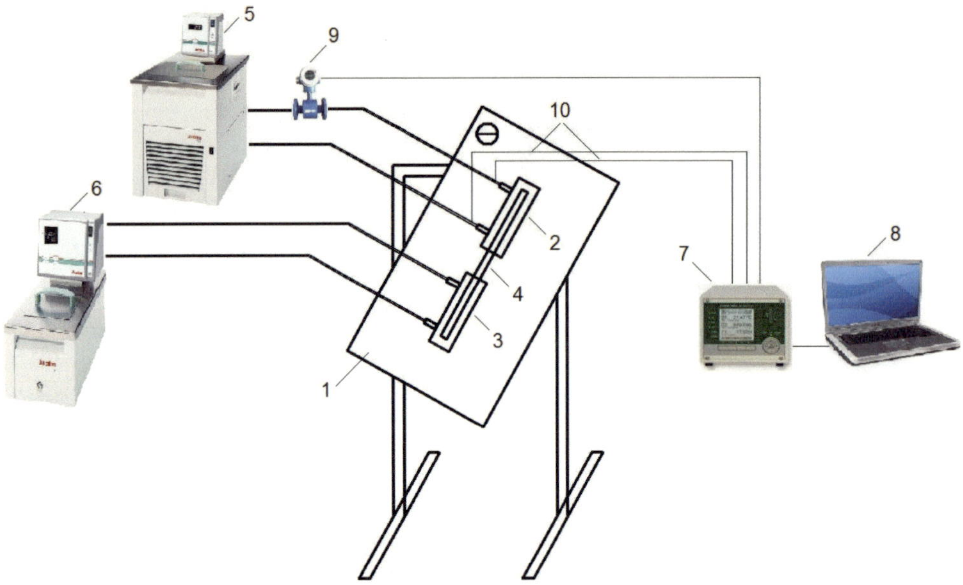

Figure 12. Scheme of measuring device: (1) rotary board, (2) cooler, (3) heater, (4) heat pipe, (5) cooling device with thermostat, (6) heating device with thermostat, (7) data logger, (8) notebook, (9) flowmeter, (10) temperature sensors.

4.1. Gravity heat pipe experiments

Gravity heat pipe experiments deal with the influence of working position, heat pipe diameter, quantity and kind of working fluid on heat transport ability gravity heat pipes. The experimental measurements were performed with heat pipe diameters (DN12, DN15, DN18, DN22 and DN28) and working fluids (distilled water, ethanol and acetone).

4.1.1. Determination of the working fluid quantity

First experiment was focused on determination optimal quantity of working fluid in heat pipe.

Determination of the quantity of the working substance is governed by basic options:

- The working fluid quantity is chosen so that the heat pipe contains both vapour and liquid over the operating temperature range.
- Lack of substance (may lead to drying of evaporator part of heat pipe).
- Surplus substance (can lead to congestion of the condensation part of the heat pipe).
- It is recommended that the working fluid to fill up at least 50% of evaporator part of the heat pipe.

This experiment was realised with heat pipe of DN 15, working fluid ethanol and working fluid quantity in range 10–50%. In **Figure 13**, results influence of working fluid quantity on heat pipe thermal performance is shown. The experiment show that the ideal quantity of

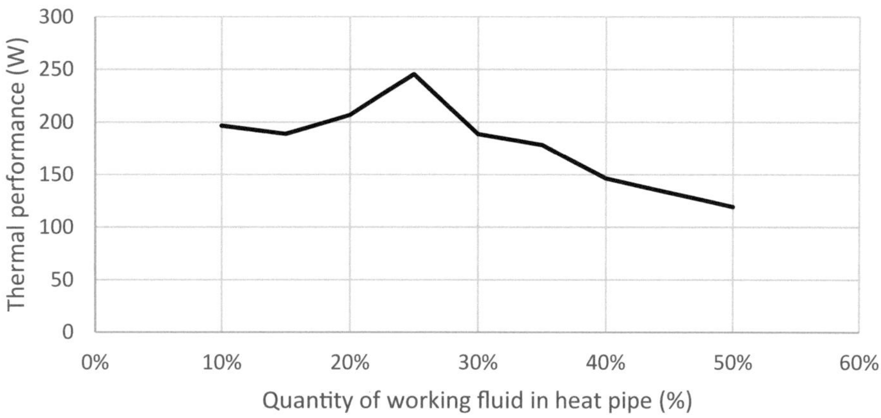

Figure 13. Influence of working fluid quantity on heat pipe thermal performance.

working fluid is in the range of 10–30% of the total heat pipe volume and the best working heat pipe was heat pipe with 25% of the total heat pipe volume from volume tube. With the increasing of working fluid quantity more than 40% the heat transfer ability rapidly decrease.

4.1.2. Influence working position on heat transfer ability of heat pipe

Second experiment was focused on the influence working position on heat transfer ability of het pipe. This experiment was realised with heat pipes of various diameters, working fluids and working positions in range from vertical to horizontal with the sequence of the inclination angle about 15°. In **Figures 14–16**,thermal performances transferred by gravity heat pipes at working positions in range 0–90° from vertical position are showed. Results show that gravity heat pipes are able transfer heat at any working position beside the horizontal position. In general, with the increasing inclination angle up to 75° from the vertical position, the heat transport ability is not changed much, even it could by say that is equal in all range or a little decrease with the top at position of 60°.

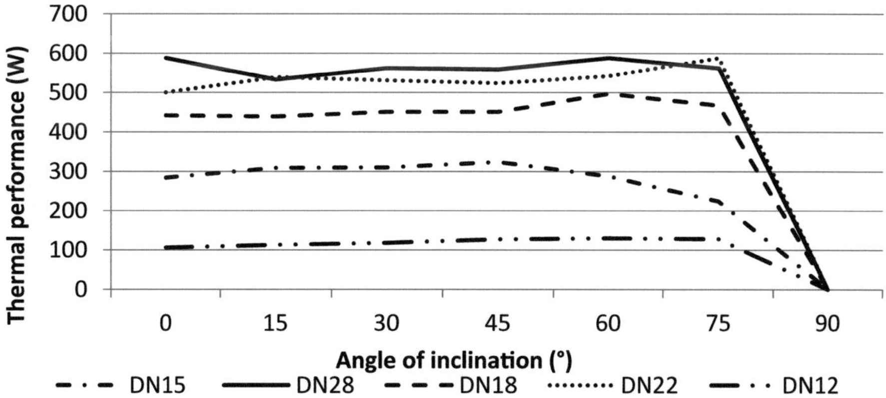

Figure 14. Influence working position on heat transfer ability of heat pipe with working fluid water.

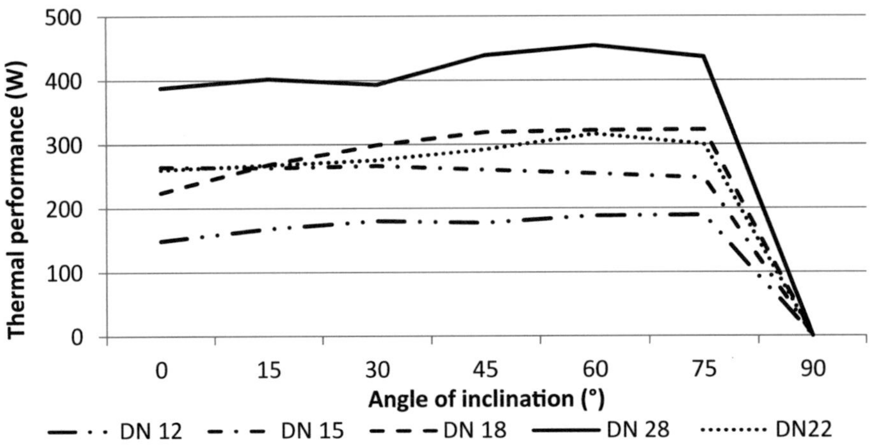

Figure 15. Influence working position on heat transfer ability of heat pipe with working fluid ethanol.

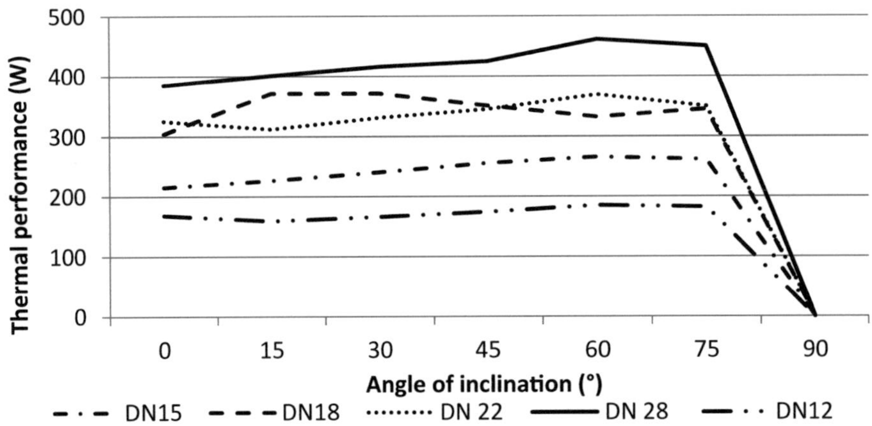

Figure 16. Influence working position on heat transfer ability of heat pipe with working fluid acetone.

4.1.3. Influence heat pipe diameter and working fluid on heat transfer ability of heat pipe

According to above experiments possible perform comparison influence heat pipe diameter and working fluid on heat transfer ability of heat pipe. In **Figure 17**, the average value of thermal performance values transferred by heat pipes with various diameter and working fluid at working position in range of 0–75° are shown. This experiment confirms assumption that with the changing heat pipe diameter and working fluid is thermal performance transferred by heat pipe changed too. With the increase of heat pipe diameter increase thermal performance.

In **Figure 18**, thermal performance transferred by heat pipe recalculated on heat flux transferred through the heat pipe cross-section. Heat flux transferred by heat pipe per cm^2 of heat pipe cross-section express expression

http://dx.doi.org/10.5772/intechopen.71543

$$Q_s = \frac{Q}{S} \tag{5}$$

where Q is thermal performance transferred by heat pipe (W), S is cross-section area of heat pipe (cm^2).

The goal of this comparison was to find, if the increase of heat pipes diameters increase heat flux transferred per cm^2. It was expect that the heat pipe with bigger diameter transferred bigger heat flux too. But this comparison show that the heat pipes with smaller diameter transferred higher heat fluxes per cm^2 of heat pipe cross-section area than heat pipes with bigger diameter.

Comparison influence of working fluid kind influence on the heat pipe heat transport ability show that the heat pipe with working fluid acetone and ethanol transfer approx. the same heat

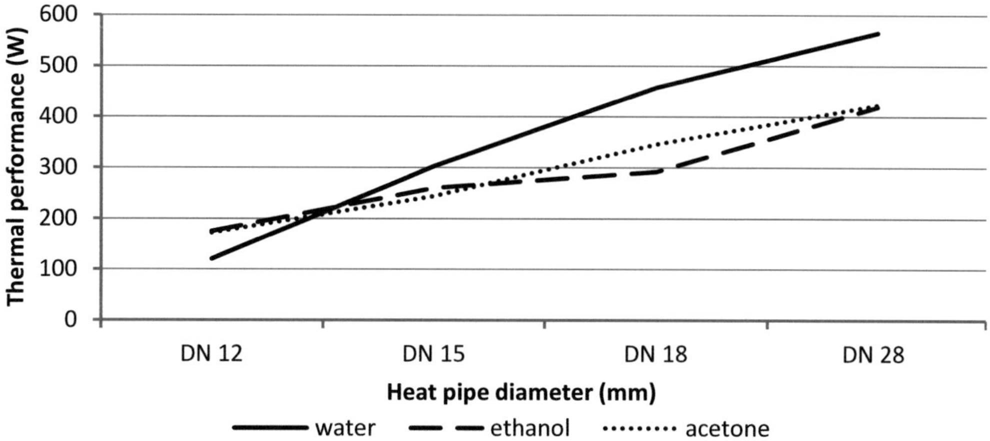

Figure 17. Dependence of heat pipe thermal performance on the working fluid and pipe diameter.

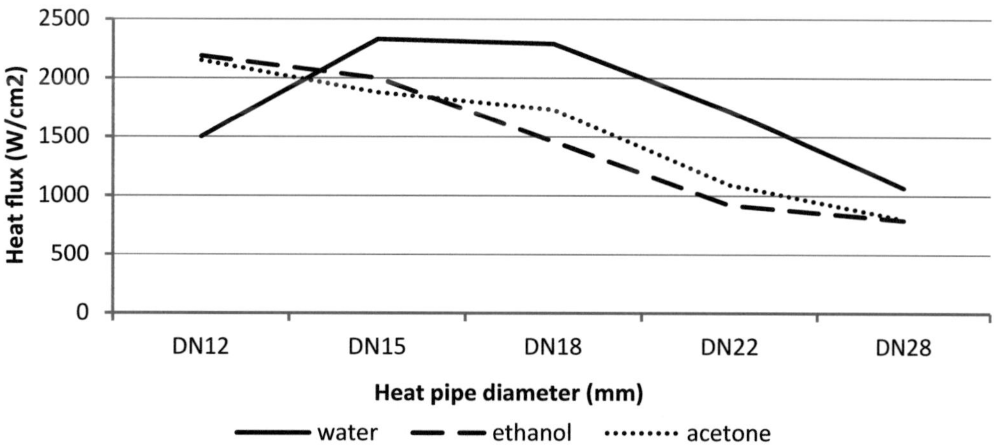

Figure 18. Dependence of heat pipe heat flux on the working fluid and pipe diameter.

fluxes. Heat pipe with working fluid water transfer higher heat fluxes than heat pipes with working fluid acetone and ethanol.

4.1.4. Working fluid flow visualisation in gravity heat pipe

Even though the theory about fluid boiling is complex described by many scientists, is hard to understand them, because boiling processes depends on many variable factors such as mass flow, vapour content and the temperature difference between fluid and the heating surface. This experiment explains working fluid flow in gravity heat pipe based on the boiling and condensation regimes and clarify phenomenon such as bubbles nucleation, liquid boiling, vapour and condensate flow interaction, vapour condensation and condensate flow down on the wall taking place inside during its operation.

4.1.4.1. Boiling regimes

In **Figure 19**, the characteristic course of boiling curve depending on the temperature difference between the evaporating working fluid and heating surface corresponding to the five known boiling regimes is shown.

Free convection evaporation: working fluid is in the liquid phase without bubbles and absorb heat transferred through the wall. The liquid close to the wall is heated and liquid at the liquid and gas interface start to evaporate.

Sub-cooled nucleate boiling region: the bubbles start nucleate in liquid phase of the working fluid due increase heat transfer between the wall and the working fluid. Created bubbles collapse in contact with the mass of liquid.

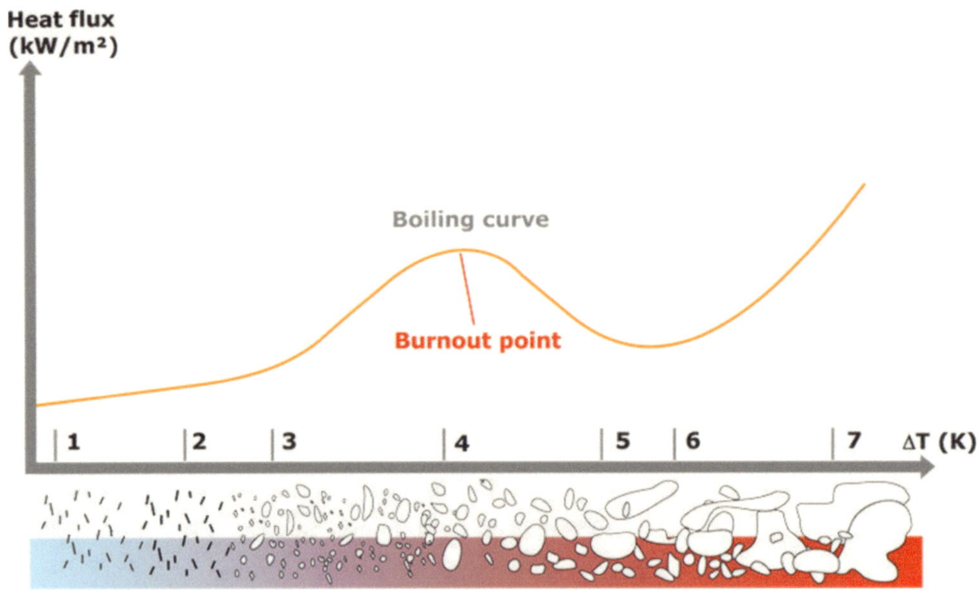

Figure 19. Boiling regimes.

Nucleate boiling region: this boiling region occur at temperature difference 3 K and is the most significant boiling region for heat transfer applications, because superheated liquid overcome surface tension of the working fluid. Increase heat transfer between wall and working fluid lead to frequent creation of bubbles which are unstable and in contact with sub-cooled liquid collapse.

Partial film boiling: further increase temperature difference cause quickly evaporation of liquid and allow another liquid access to the heated surface. This is unstable boiling regime and is not eligible, because vapour bubbles are create more frequent which act as an insulation and it cause heat transfer reduction.

Complete film boiling: at very high temperature difference between working fluid and wall a stable film of working fluid vapour on the wall is creating. Even though the film act as an effective insulation and heat transfer is minimal is this boiling region much more preferable than unstable partial film boiling.

4.1.4.2. *Condensation regimes*

Condensation is phase change of the working fluid from the vapour to the liquid. It occurs in condensation section on the wall heat pipe and is accompanied by heat and mass transfer. During condensation latent heat is released and thus increases the heat pipe efficiency.

In **Figure 20**, two types of condensation occur on the heat pipe wall are shown.

Film condensation: occurs when the working fluid condensate in form of continuous thin layer (film) liquid on the surface wall. The liquid film flows down the wall to the evaporator section due gravity and thus extend heat transfer.

Drop-wise condensation: occurs when the working fluid condensate in form of small liquid droplets on the surface wall. With the increasing vapour condensation, droplets grown and collect to the bigger drops until due gravity action flow down the wall to the evaporator. A new drop-wise condensation can proceed on free space after flowing down drops.

Even though the drop-wise condensation has 10 times larger heat transfer coefficient than film condensation, it is difficult usable in practice due short duration. In view, the heat transfer

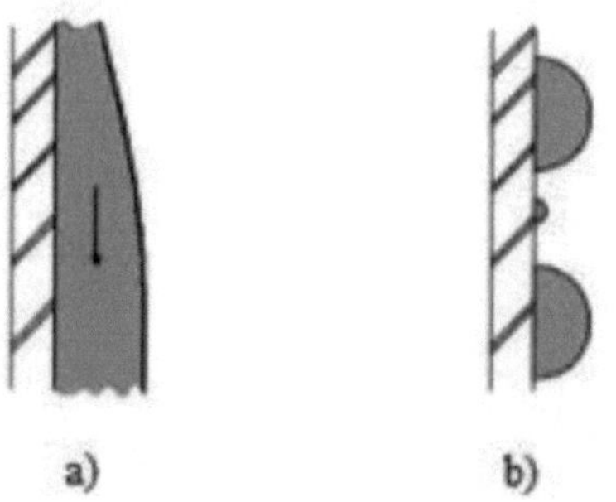

Figure 20. Condensation regimes: (a) film condensation, (b) drop-wise condensation.

coefficient is the continuous drop-wise condensation and film condensation with thin layer the most desirable condensation regimes in heat pipe.

The flow visualisation of the working fluid in gravity heat pipes was performed on two heat pipes made of glass with various inner diameter 13 and 22 mm. The total length of heat pipes is 500 mm and water was used as working fluid. Experiment consists of scanning the working fluid flow inside the glass heat pipe during its operation by high speed video camera. The results of the working fluid flow visualisation can observe boiling and condensation take place in heat pipe and can be helpful in two-phase flow or heat and mass transfer simulation by CFD method.

In **Figure 21**, heat pipes made for the experiment and experiment process of the working fluid flow visualisation in gravity heat pipe are shown. The heat pipe construction consists of borosilicate glass container, Cu cap on the bottom and brass filling valve on the top.

The borosilicate glass container of heat pipe was chosen because has smooth surface, ensure good visibility and is high temperature, chemical and water resistant. The heat pipes were manufactured by working fluid evaporation method. This is the simplest method how to make a heat pipe. At first is total heat pipe volume filled with working fluid and then is working fluid evaporated until the required volume. Thus it ensures only pure liquid and vapour phase of working fluid and no undesirable gases inside the heat pipe.

The results of the working fluid flow visualisation in heat pipe are shown in **Figure 22**. In **Figure 22a** and **b**, working fluid flow regimes in evaporator and condenser section of gravity heat pipe with inner diameter 13 mm are shown.

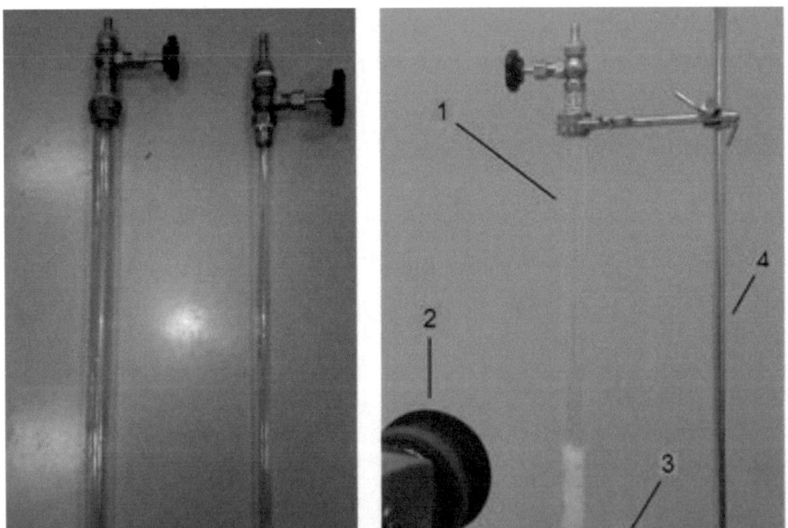

Figure 21. Working fluid flow visualisation in gravity heat pipe: (1) heat pipe, (2) video camera, (3) heat source, (4) stand.

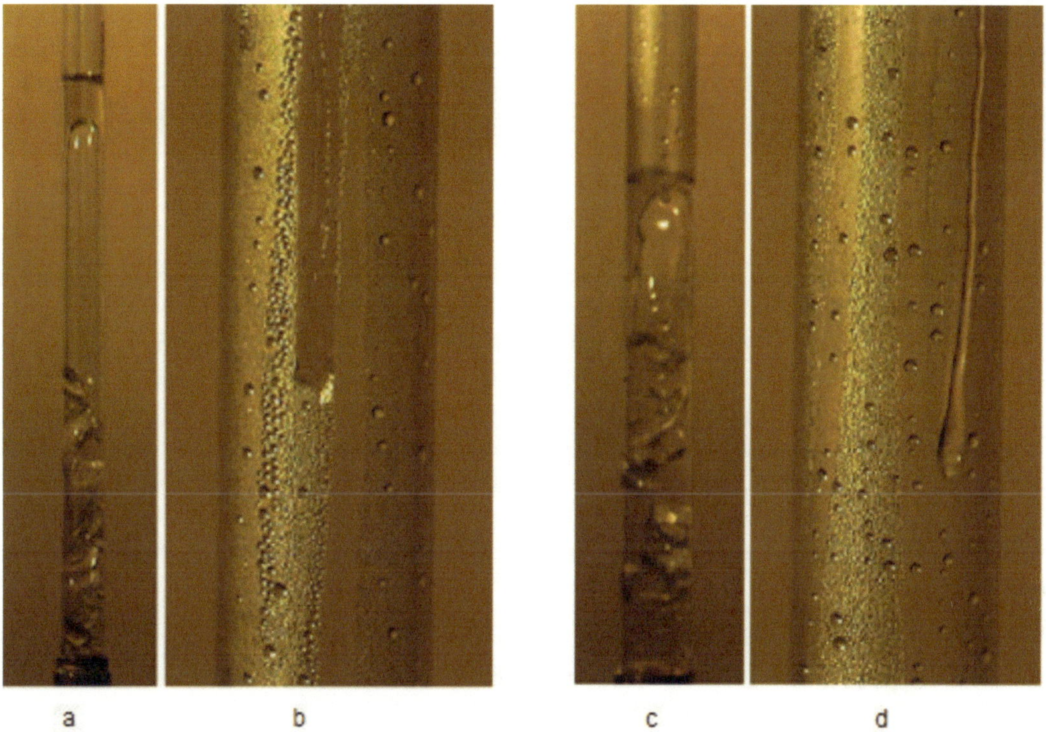

Figure 22. Boiling and condensation regimes of glass gravity heat pipe with working fluid water. (a) Evaporation section of heat pipe with inner diameter 13 mm, (b) Condensation section of heat pipe with inner diameter 13 mm, (c) Evaporation section of heat pipe with inner diameter 22 mm, (d) Condensation part of heat pipe with inner diameter 22 mm.

There you can see a typical film boiling regime with some big vapour bubbles on the bottom of heat pipe evaporator and one big vapour blanket push out liquid to the adiabatic section. In the condenser section you can see a typical drop-wise condensation when a small droplets of the working fluid are created on the surface wall. Once a time they are collect to the big drops, flown down the wall to the evaporator due to gravity and create a free space for a new drops.

In **Figure 22c** and **d**, working fluid flow regimes in evaporator and condenser section of gravity heat pipe with inner diameter 22 mm are shown. There you can show a similar film boiling regime as in heat pipe with the inner diameter 13 mm with the some big vapour bubbles on bottom of heat pipe evaporator In this case vapour did not push out liquid to the adiabatic section so high, what may a bigger heat pipe diameter cause. In the condenser section, you can see the same drop-wise condensation of the working fluid on the surface wall as in heat pipe with the inner diameter 13 mm. There was not observed influence of the heat pipe diameter on the condensation regime.

4.2. Wick heat pipe experiments

Wick heat pipe experiments deal with the influence of working position and quantity of working fluid on heat transport ability wick heat pipes. The experimental measurements was performed with one heat pipes with grooved wick structures, three heat pipes with mesh screen wick structure (mesh 50,100 and 200), three heat pipes with porous sintered wick structure (metal powder 35 , 63 and 100 μm) and one heat pipe with composite wick structure (groove + mesh screen 50). All heat pipe has diameter DN 15, total length 0.5 m and working fluid water.

4.2.1. Determination of the working fluid quantity

At first was performed experiment focused on determination optimal quantity of working fluid in heat pipe. To determination optimal quantity of working fluid in wick heat pipe apply the same rules as in gravity heat pipe. This experiment was realised with heat pipe with mesh screen wick structure of DN 15, working fluid water and working fluid quantity in range 10 to 50%. In **Figure 23**, results influencing of working fluid quantity on heat pipe thermal performance at various working position are shown . The working position can be divided into three zones. Positive gravity action zone is represent by angle of inclination from vertical position 0–75°, zero gravity action zone (horizontal position) is represent by angle of inclination from vertical position 90° and negative gravity action zone is represent by angle of inclination from vertical position 105–180°.

In **Figure 24** results influencing working fluid quantity on heat pipe thermal performance at horizontal position are shown. This experiment show that the ideal quantity of working fluid is in the range of 10–30% of the total heat pipe volume and the best working heat pipe is heat pipe with 25% of the total heat pipe volume. Wick heat pipe with 40% is good operating in the gravity positive and zero action zone but in the negative gravity action zone. Wick heat pipe with working fluid quantity 50% and more are does not able heat transfer in zero and negative gravity action zone.

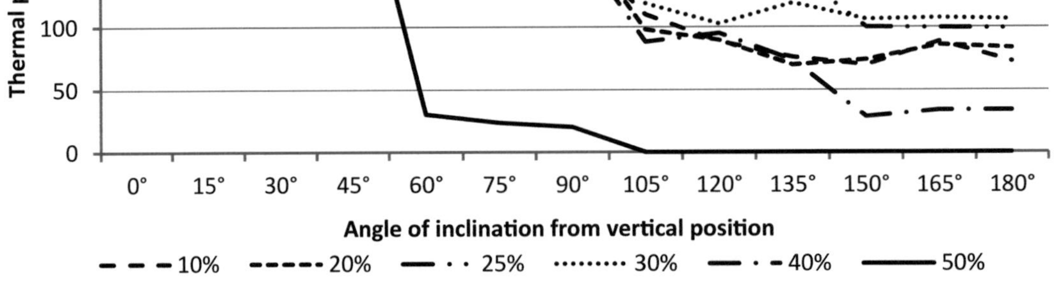

Figure 23. Dependence thermal performance on working position and various quantity of working fluid of wick heat pipe with mesh screen wick structure and water working fluid.

4.2.2. Influence working position on the heat transfer ability of heat pipe

In **Figure 25** influence of working position on the heat transfer ability of heat pipe with various wick structures (groove, mesh screen, sintered and composite) are shown. Working position of heat pipe can divided into three areas. Positive gravity action zone is represent by angle of inclination from vertical position 0–75°, zero gravity action zone (horizontal position) is represent by angle of inclination from vertical position 90° and negative gravity action zone is

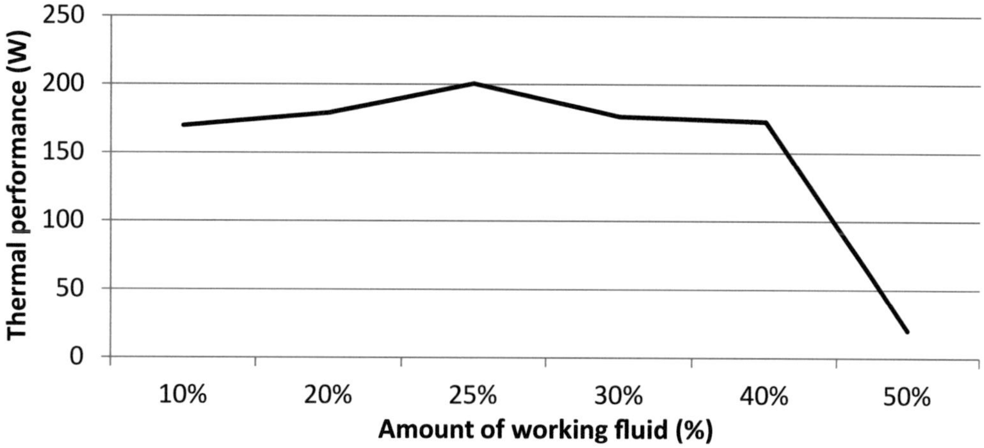

Figure 24. Dependence thermal performance on working fluid amount of wick heat pipe with mesh screen wick structure and water working fluid at horizontal position.

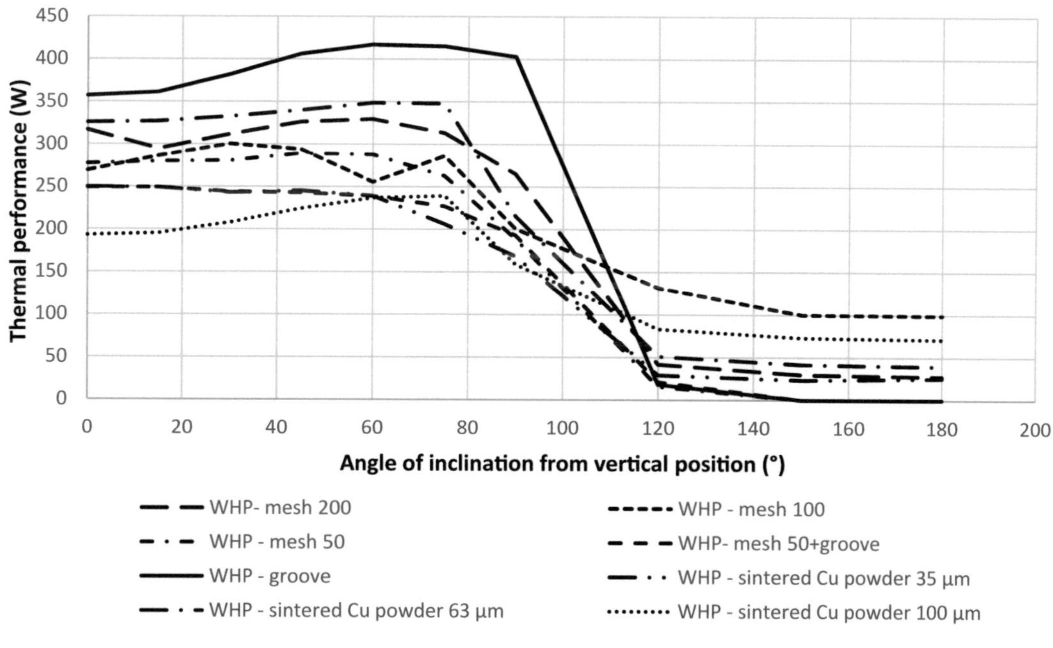

Figure 25. Dependence thermal performance on working position of wick heat pipes with various wick structures.

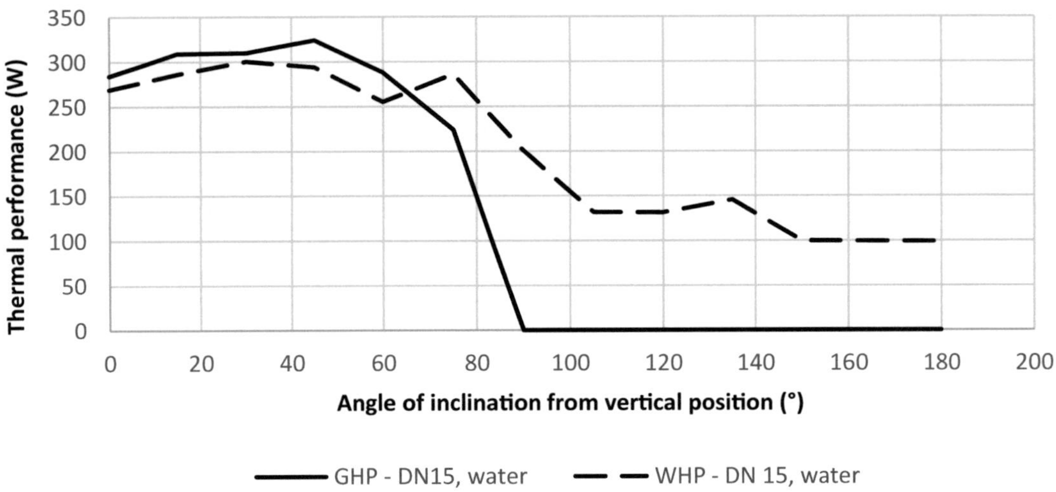

Figure 26. Dependence thermal performance on working position of gravitation heat pipe and wick heat pipe with mesh screen wick structure.

represent by angle of inclination from vertical position 105–180°. There is seen that all WHP has good ability heat transfer in positive and zero gravity action zone. The best ability shows WHP with groove wick structure. WHP with mesh screen wick structure show better ability than WHP with porous sintered wick structure. Zone with negative gravity action shows good ability heat transfer WHP with mesh and sintered wick structure. The best working heat pipes are WHP with wick structure mesh 100 and sintered 100 μm. WHP with groove wick structure a mesh 50 wick structure does not transfer heat. WHP with composite wick structure (mesh 50 + groove) show good ability heat transfer in zone of positive and zero gravity action but in the zone of negative gravity action does not heat transfer.

4.2.3. Influence of the working position on the heat transfer ability

In **Figure 26**, courses thermal performances transferred by WHP and GHP depending on working position are shown. At working position with positive gravity action in the region inclination angle from vertical position of 0–75° are courses of WHP and GHP are approx. similar when GHP is able transfer higher thermal performance than WHP. At the horizontal position when is the gravity action zero and in the vertical position with evaporator section above condensation section when is the gravity action negative are see differences when WHP is able transfer heat due the wick structure and GHP does not able transfer heat.

5. Heat pipe calculation

The heat flux transferred through the heat pipe depends mainly on the temperature difference and the corresponding thermal resistances. The real transferred heat is affected by the hydrodynamic and thermal processes that take place in the heat pipe at the various operating conditions. The heat flux transferred by the heat pipe can reach limit values that depend on these processes. There are five known limitations that limit the overall heat transfer in different parts of the heat pipe depending on the working temperature. In **Figure 27**, an ideal model of all heat transfer limitations that define area of maximum heat flux transferred by heat pipe depending on operating temperature is shown.

5.1. Mathematical model

The mathematical model consist of calculation the heat pipe heat transfer limitation calculation. Heat pipe heat transfer limitations depending on the working fluid, the wick structure, the dimensions of the heat pipe and the heat pipe operation temperature. Each heat transfer limitation express part of total heat flux heat pipe, which is influencing hydrodynamic and thermal processes that occur in the heat pipe. Each limitations exists alone and they are oneself non-influence together. To design mathematical model for calculating heat flux transferred by heat pipe is necessary to know basic and derived parameters of the heat pipe and its wick structure and physical properties of the working fluid liquid and vapour phase.

5.1.1. Capillary limitation

Capillary limitation involves a limitation affect the wick heat pipe operation, which results from the capillary pressure acting on the condensed working fluid in the capillary structure. At the contact of liquid and wick structure surface, the capillary pressure is formed. This causes the liquid phase of the working fluid flow from the condenser to the evaporator. Decreasing the pores of the capillary structure increases the capillary pressure as well as hydraulic resistance. The capillary limit occurs when the capillary forces at the interface of the liquid and vapour phases in the evaporator and condenser section of the heat pipe are not large enough to overcome the pressure losses generated by the friction. If the capillary pressure in the heat pipe

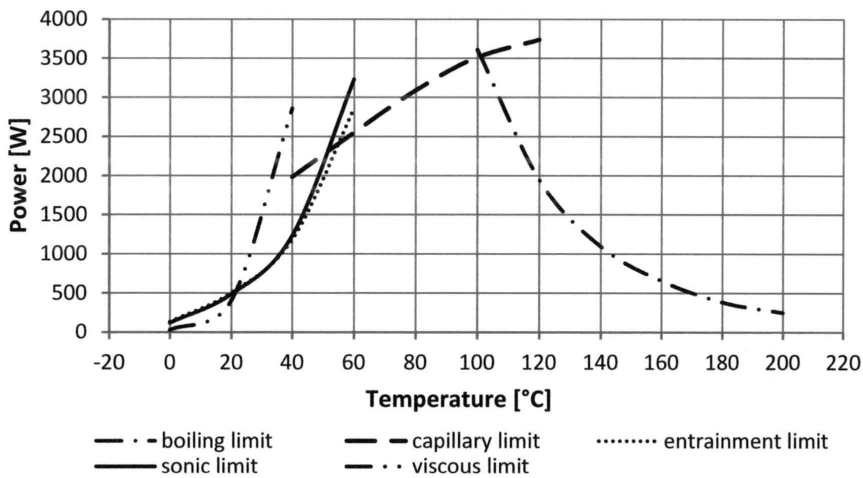

Figure 27. Heat transfer limitations of water wick heat pipe with sintered wick structure (heat pipe inner diameter 20 mm, total length 2 m, axial orientation 90°, sphere diameter of copper powder 0.85 mm, porosity 0.55, width of wick structure 6 mm).

during the operation is insufficient to provide the necessary condensate flow from the condenser to the evaporator, the capillary structure in the evaporator is dried and thus the further evaporation of the working substance is stopped. In general, the capillary limit is the primary limit that influences the heat pipe performance and is expressed by the relationship [12].

$$\dot{Q}_c = \frac{\sigma_l.\rho_l.l_v}{\mu_l}.\frac{K.A_w}{l_{eff}}.\left(\frac{2}{r_{eff}} - \frac{\rho_l.g.l_t.\cos\Psi}{\sigma_l}\right) \tag{6}$$

where A_w is the wick cross-sectional area (m^2), K is the wick permeability (m^2), μ_l is the liquid viscosity ($N.secm^{-2}$), ϱ_l is the liquid density ($kg\ m^{-3}$), g is the acceleration due to gravity ($9.8\ m\ sec^{-2}$), r_{eff} is the wick capillary radius in the evaporator (m) and l_t is the total length of the pipe (m) [13].

Furthermore, if the heat pipe has properly operate, the maximum capillary pressure have to be greater than the total pressure loss in the heat pipe and it is expressed by the relationship

$$(\Delta P_c)_{max} \geq \Delta P_{tot} \tag{7}$$

The maximum capillary pressure ΔP_c developed in wick structure of the heat pipe is defined by the Laplace-Young equation.

$$\Delta P_c = \frac{2\sigma}{r_{eff}}.\cos\,\theta. \tag{8}$$

where r_{eff} is the effective pores radius of the wick structure and θ is contact angle liquid phase of the working fluid in wick structure, where $\theta = 0°$ is the best wetting contact angle [5].

5.1.2. Viscous limitation

When the heat pipe operate at low operating temperatures the saturated vapour pressure may be very small and has the same range as the required pressure drop necessary to vapour flow from the evaporator to the condenser of the heat pipe. This results in a condition expressed by the viscous limit about balance of the vapour pressure and viscous forces in the capillary structure in the low velocity vapour flow. The most frequent cases of exceeding the boundary of the viscous limit occur when the heat pipe operate at temperature close the solidification of the working fluid. In this case, working fluid evaporation in the evaporator and heat transfer in the form of vapour flow through the adiabatic section into the condenser of the heat pipe did not occur. It is assumed that the vapour is isothermal ideal gas, the water vapour pressure on the end of the condenser is equal to zero, which provides the absolute limit for the pressure in the condenser. The viscous limit is referred as the condition of the vapour phase flow at low velocity and is expressed by the relationship

$$Q_v = \frac{\pi.r_v^4.l_v.\rho_v.P_v}{12.\mu_v.l_{eff}} \tag{9}$$

where l_v is the latent heat of vaporisation (J/kg), r_v is the cross-sectional radius of the vapour core (m), l_{eff} is the effective length of the heat pipe (m), μ_v is the vapour viscosity in the evaporator ($N.secm^{-2}$), P_v (Pa) is the vapour pressure and ϱ_v (kgm^{-3}) is the density at the end of the heat pipe evaporator [14].

In case when the viscous limit is reached for many conditions, the condenser pressure could not be a zero. Then the following expression is applied

$$Q_v = \frac{A_v.r_v^2.l_v.\rho_v.P_v}{16.\mu_v.l_{eff}}.\left(1 - \frac{P_{v,c}^2}{P_v^2}\right) \tag{10}$$

where $P_{v,c}$ is the vapour pressure in the condenser.

5.1.3. Sonic limitation

The sonic limit characterises the state in which the velocity of the evaporated vapour flow at the outlet of the evaporator reaches the sound velocity. Generally, this phenomenon occurs on the start of heat pipe operation at a low vapour pressure of the working fluid. Assuming that the vapour of the working fluid is the ideal gas and the vapour flow at the sound velocity throughout the heat pipe cross-section is uniform, the sonic limit is determined by the relationship (11). The sonic limit does not depend on the heat pipe orientation, type of the heat pipe and the same formula is applied for the gravity and wick heat pipe. The most difficult in the sonic limit determination is determining quantities of vapour density and pressure on inlet to the condenser [15].

$$Q_s = 0,474.A_v.l_v.\left(\rho_v.P_v\right)^{0,5} \tag{11}$$

where ϱ_v (kgm^{-3}) is the vapour density, P_v (Pa) is pressure at the end of heat pipe evaporator and A_v is the cross-sectional area of the vapour core (m^2).

The sonic limit is mainly associated with liquid-metal heat pipe start-up or low-temperature heat pipe operation due the very low vapour densities that occur in these cases. For the low-temperature or cryogenic temperatures, the sonic limit is not a typically factor, except for heat pipes with very small vapour channel diameters. The sonic limitation is referred as an upper limit of the axial heat transport capacity and does not necessarily result in dry-out of the wick structure in heat pipe evaporator or total heat pipe failure.

5.1.4. Entrainment limitation

Increasing the heat flux transferred by heat pipe increases the vapour flow velocity of the working fluid increases too and this result to a more pronounced interaction of the vapour and liquid phases inside the heat pipe. The interfacial surface becomes unstable and the viscous forces on the surface of the liquid overcome the forces of the surface tension. The wave are creates on the liquid phase surface at first from which the droplets are gradually tearing off. At a certain vapour flow velocity the liquid flow interruption into the evaporator section occur. The condenser section of heat pipe is overflow by vapour and liquid phase and the evaporator is overheated due to lack of the working fluid. The limit value of the heat flux at which is the heat pipe overflow corresponds to interaction limit. The entrainment limit calculation of the gravity heat pipe is based on the empirical correlation of a counter current air and water film flow in a vertical tube. Entrainment limitation in gravitational heat pipe occur when the velocity of the liquid film approximates to zero and could by expressed by relationship [16]

$$Q_i = C^{x^2}.A_v.l_v.\left[\rho_l^{\frac{-1}{4}} + \rho_v^{\frac{-1}{4}}\right]^{-2}.\left[g\sigma_l.(\rho_l - \rho_v)\right]^{\frac{1}{4}}, \tag{12}$$

$$C^{x^2} = 1,79tgh\left(0,5Bo^{\frac{1}{4}}\right), \tag{13}$$

$$Bo = d_p\left[\frac{g(\rho_l - \rho_v)}{\sigma_l}\right]^{-2} \tag{14}$$

Entrainment limitation of the wick heat pipe is related to the condition when the vapour flows against the liquid flow in the wick structure, which may result in insufficient liquid flow in the wick structure [17]. Entrainment limitation of the wick heat pipe is expressed by relationship

$$Q_e = A_v.l_v.\left(\frac{\rho_v.\sigma_l}{2.r_{c,\,ave}}\right)^{0.5} \tag{15}$$

where $r_{c,ave}$ is the average capillary radius of the wick structure and in many cases is approximated to r_{eff}, σ_l is the liquid surface tension (N/m).

5.1.5. Boiling limitation

When heating the surface of the heat pipe wall with a layer of liquid in the saturation boundary a three basic heat transfer regimes can occur. At low-temperature difference of the heated surface and interfacial surface of the liquid, a natural convection and evaporation from the liquid surface occurs. When increasing the temperature difference, a bubble boiling and gradually transformation to the film boiling occur. In heat pipe a surface evaporation at low heat flux densities and bubble boiling at higher densities occur. Although the heat transfer intensity is greatest in the bubble boiling, for most types of wick heat pipes the bubble boiling is not desired because interfere with the liquid wicking into the wick structure. On the other hand, in a heat pipe with a grooved capillary structure and a gravity heat pipe is bubble boiling favourable [18]. The heat flux in which the bubble boiling occurs in the wick heat pipes and the film boiling occurs in the gravity heat pipe is referred as the boiling limit. For the gravity heat pipe is expressed by the relationship [19]

$$Q_v = 0.16.A_v.l_v\sqrt[4]{\sigma_l.g.\rho_v^2.(\rho_l - \rho_v)} \tag{16}$$

Determination of the boiling limit of the wick heat pipe is problematic, because it depends on a number of technological and operating conditions. The most reliable determination of the boiling limit is experimental determination for the particular wick structure and working fluid. Approximate determination of the boiling limitation for the wick heat pipe is expressed by the relationship [20]

$$Q_b = \frac{4\pi.l_{eff}.\lambda_{eff}.T_v.\sigma_l}{l_v.\rho_v.\ln\frac{r_i}{r_v}}.\left(\frac{1}{r_n} - \frac{1}{r_{eff}}\right) \quad (17)$$

where λ_{eff} is the effective thermal conductivity of the wick structure compose of the wick thermal conductivity and working fluid thermal conductivity (W/m K), T_v is temperature of vapour saturation (K), r_v is the vapour core radius, r_i is the inner container radius (m) and r_n is the bubble nucleation radius in range from 0.1 to 25.0 μm for conventional metallic heat pipe container materials.

5.1.6. Heat pipe parameters

To calculation heat pipe heat transport limitations is need to know thermophysical properties of working fluid in heat pipe, basic heat pipe parameters, thermal conductivity of heat pipe material, working temperature of heat pipe, axial orientation of heat pipe and others heat pipe parameters calculated from basic heat pipe parameters needed.

$$l_t = l_e + l_{ad} + l_c \quad (18)$$

$$l_{eff} = 0,5.(l_e + l_c) + l_{ad} \quad (19)$$

$$A_v = \pi.r_v^2 \quad (20)$$

$$A_w = \pi.\left(r_i^2 - (r_i - h)^2\right) \quad (21)$$

where l_t is total length of heat pipe (m), l_e is evaporation length of heat pipe (m), l_{ad} adiabatic length of heat pipe (m), l_c is condensation length of heat pipe (m), l_{eff} is effective length of heat pipe (m), A_v is cross-sectional area of the vapour core (m^2), A_w is wick cross-sectional area (m^2), r_v is cross-sectional radius of vapour core (m), r_i is inner container radius (m) and h is wick structure width (m).

The other parameters needed to calculation heat pipe heat transport limitations are basic parameters of sintered wick structure and others parameters calculated from basic parameters of wick structure.

$$r_{eff} = 0,21.d_s \quad (22)$$

$$K = \frac{d^2.\varepsilon^3}{150.(1-\varepsilon)^2} \quad (23)$$

$$\lambda_{eff} = \lambda_l \frac{2.\lambda_l + \lambda_m - 2.(1-\varepsilon).(\lambda_l - \lambda_m)}{2.\lambda_l + \lambda_m + (1-\varepsilon).(\lambda_l - \lambda_m)} \quad (24)$$

where K is permeability (m^2), d is sphere diameter (m), ε is porosity (−), r_{eff} is effective radius of wick structure (m), λ_{eff} is effective thermal conductivity, λ_l is thermal conductivity of working fluid liquid and λ_m is thermal conductivity of wick material [21].

5.2. Verification of the mathematical model

The mathematical model was created according above equations of limitations and input heat pipe parameters. Results of mathematical model are graphic dependencies of heat transport limitations on heat pipe working temperature. Mathematical model results of heat transport limitations specific types heat pipe was compare with results from measurement of heat pipe performance at temperature 50 and 70°C. In **Figure 28** are graphic comparison results heat transport limitations determining total performance of heat pipe from mathematical model with measured performance of ethanol wick heat pipe with sintered wick structure and sphere diameter of copper powder 0.1 mm. Dotted line create boundary of heat pipe performance by capillary limitation and dashed line is boiling limitations. The full line is measured results of heat pipe thermal performance at temperature 50 and 70°C. In **Figure 28** is seen that the dotted line and full line are in the same region at temperature 50 and 70°C.

Results in the next **Figure 29** confirm verification of mathematical model, where measuring heat flux of wick heat pipe with grooved wick structure is in similar region as a calculation

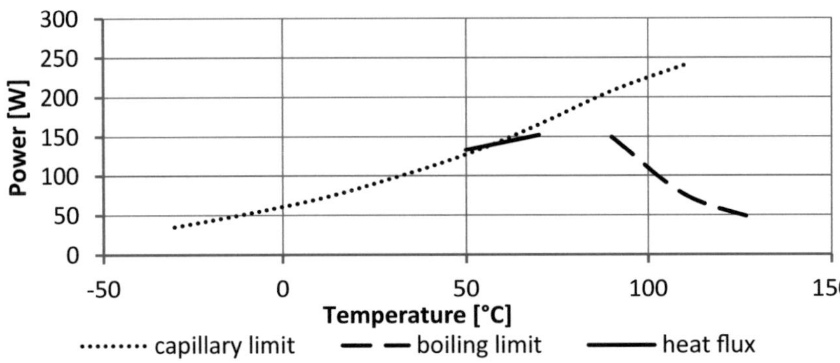

Figure 28. Verification of mathematical model by measuring of heat pipe performance (ethanol wick heat pipe with sintered wick structure and sphere diameter of copper powder 0.1 mm and axial orientation of heat pipe ψ 180°).

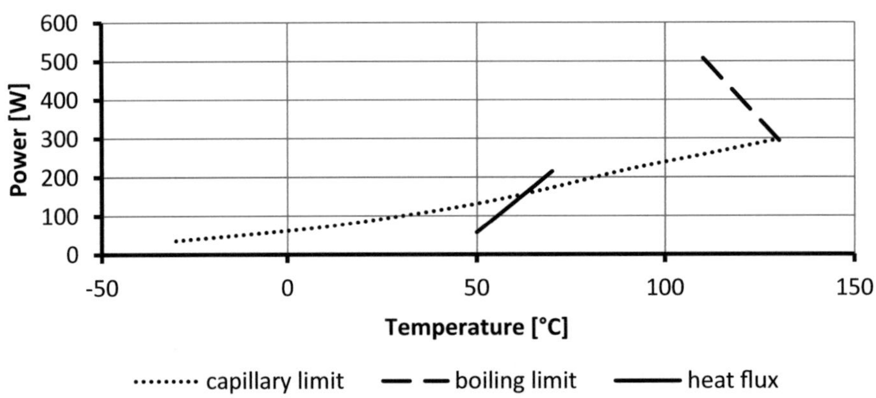

Figure 29. Verification of mathematical model by measuring of heat pipe performance (ethanol wick heat pipe with grooved wick structure length 0.3/width 0.2/pitch 0.3).

results of capillary limitation, which is determining limitation for this kind of heat pipe at temperature 50 and 70°C. There is seen that the capillary limitation is determining limitation at full working temperature range.

5.3. Results of heat pipe calculation

Results of the heat pipe calculation show interesting graphs of the maximal heat flux transferred by heat pipe depending on the wick structure parameters. It could be used in design optimization of the heat pipe wick structure. The curves present area of maximal heat flux transferred by heat pipe depends on operating temperature. In **Figures 30** and **31**, influence of groove dimensions on total heat pipe performance is presented. Heat transport limitations of wick heat pipe with grooved wick structure and groove dimensions (height 0.3 mm, width 0.2 mm and pitch 0.3 mm) created by mathematical model are presented. In **Figure 30**, influence of groove high from 0.3 to 0.9 mm on total heat pipe performance is shown. Heat pipe performance rising with groove high increase was seen. But increasing of groove height from 0.7 to 0.9 is the boiling limitation shown as a main limitation and at working temperature from 80 to 130°C has decrease tendency on heat pipe performance. According to graphic dependencies, the groove height of 0.6 is shown as an optimal height for this specific type of grooved wick heat pipe.

In **Figure 31**, influence of groove width from 0.2 to 0.9 mm on total heat pipe performance is shown. An extreme increase of heat pipe performance at temperature range from 40 to 90°C is seen. These temperature range determines the limitation a capillary limitation. At lower temperature of working temperature range is with increasing of groove width the main limitation sonic limitation. At higher temperature, boiling limitation as a main limitation and for groove

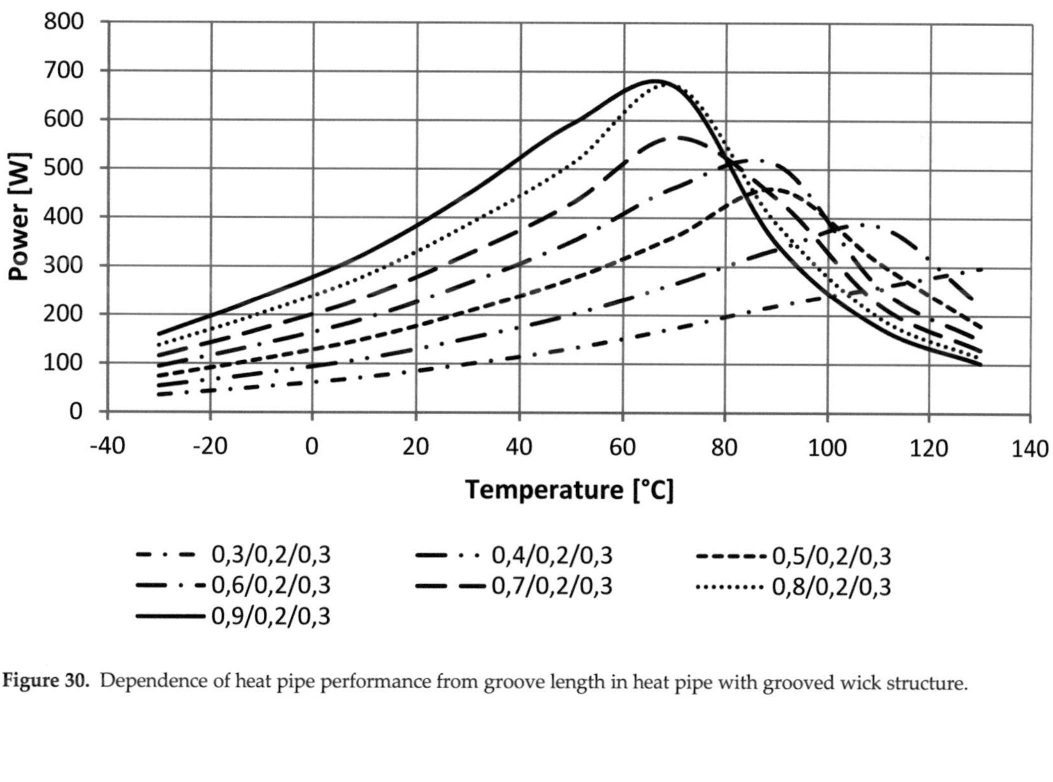

Figure 30. Dependence of heat pipe performance from groove length in heat pipe with grooved wick structure.

width is shown in the range of 0.6–0.9 for boiling limitation equal at temperature from 80 to 130°C. According to graphic dependencies, the groove width of 0.6 is shown as an optimal width for this specific type of grooved wick heat pipe.

In **Figure 32**, influence of wick structure width on heat pipe performance is shown. Wick structure width is an important factor, which influences heat pipe performance. There is seen that the heat pipe performance increase with the wick structure thickness in operating temperature region of −30 to 60°C. The capillary limitation is a main limitation for this region. On the other way, increase of the wick structure thickness decrease the heat pipe performance in

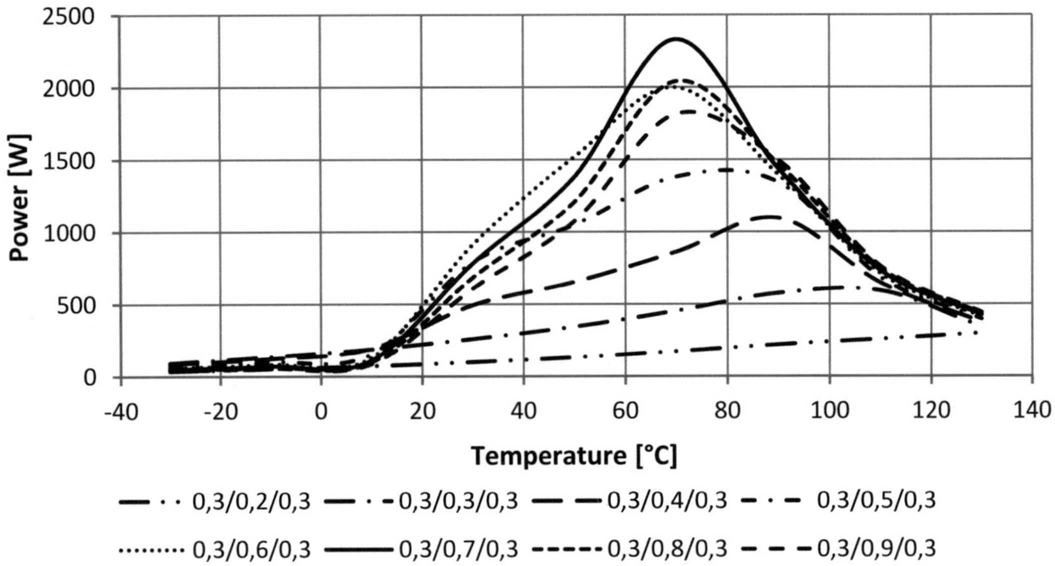

Figure 31. Dependence of heat pipe performance from groove width in heat pipe with grooved wick structure.

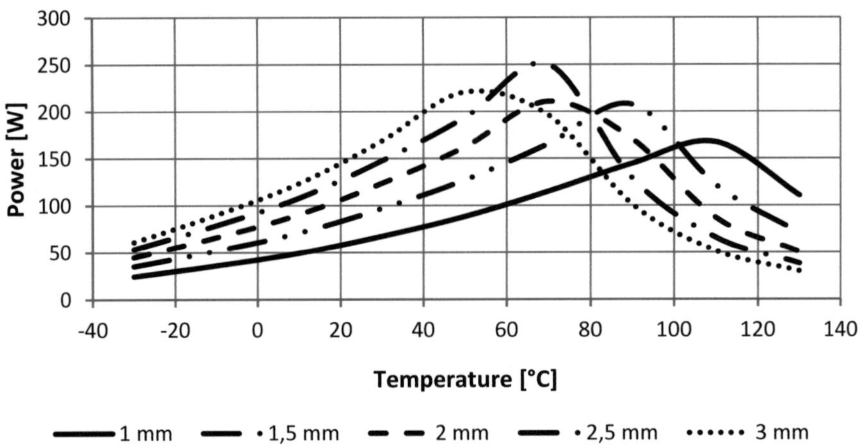

Figure 32. Dependence of heat pipe performance from wick structure width of the sintered wick heat pipe.

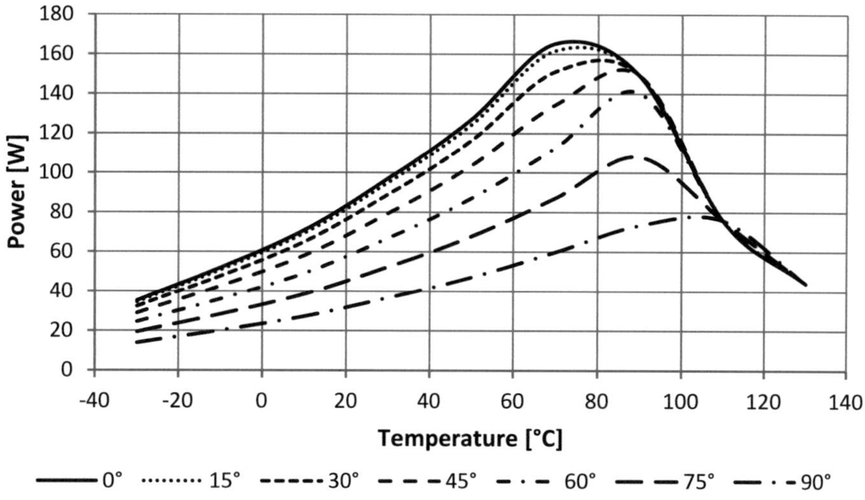

Figure 33. Dependence of heat pipe performance from position of the sintered wick heat pipe.

operating temperature region of 80–130°C. It may be caused by bubble nucleation in wick structure, when the returning liquid from the condenser section to evaporator section of heat pipe evaporates. In this case, the main limitation is boiling limitation.

In **Figure 33**, influence of the position heat pipe on their performance is shown. Wick in heat pipe ensure liquid return from condensation section to evaporation section of heat pipe and therefore wick heat pipe can operate at various tilt angles, even at horizontal position. The heat pipe performance decrease with higher tilt angle from vertical position. But even though according graphic dependencies of heat pipe performance on heat pipe position, it can say that heat pipe performance of wick heat pipe in horizontal position is only at half less as at vertical position.

6. Conclusion

The experiments performed with the heat pipes in this work gives several conclusions about influence working position on their heat transfer ability, where the gravity and capillarity of wick structures plays main role.

The results of performed experiments show how the gravity effect on the heat transport ability of gravity and wick heat pipe type at changing working position. Gravity heat pipe can operate only in working position with positive action of gravity. The heat transport ability of GHP with the change in working position from vertical (0°) to horizontal (90°) is changed too. There is an interesting finding that heat transport ability of gravity heat pipe with the increasing inclination angle up to 75° from the vertical position, does not change much even it could say that in the position from 0 to 60° slightly increase. With next increasing inclination angle from 75 to 90°, the heat transport ability of gravity heat pipe rapidly decrease to zero value. Wick heat pipe can operate in positive, zero and negative action of gravity, while the heat transfer ability of the wick heat pipe is uniform in area of the positive and zero action gravity.

In area of the negative action of gravity, the heat transfer ability of wick heat pipe gradually decrease, but still is able to transfer heat.

The other conclusion of this work is that with increasing heat pipe diameter, the thermal performance transferred by gravity heat pipe is increasing too. But this statement does not apply for the heat flux transferred per cm^2 of heat pipe cross-section area, because heat pipes with smaller diameter transferred higher heat fluxes per cm^2 of heat pipe cross-section area than heat pipes with bigger diameter.

Experiment of working fluid flow visualisation in gravity heat pipe show that the inner diameter of heat pipe does not have influence on the boiling and condensation regimes. In both cases, the film boiling in the evaporator section and drop-wise condensation in condenser section occur. The boiling regimes differ only in height of pushed liquid in to the adiabatic section. The pushed liquid height was in heat pipe of inner diameter 13 mm higher than in the heat pipe of inner diameter 22 mm. In both cases of drop-wise condensation, the small drops collect to the bigger drop and flow down the wall due gravity.

The mathematical calculation of the heat pipe, heat transport limitations show that the critical limitations influencing heat transfer ability of wick heat pipe are entrainment limitation, capillary limitation and boiling limitation. These limitations depends on thermophysical properties, wick and heat pipe parameters. The thermophysical properties of each working fluid are stable in temperature range and they cannot change. Changing the dimensions of wick structure is possible that optimise total heat flux transferred by heat pipe, because capillary pressure makes the wick structure depend mainly on the wick structure permeability. When design wick structure, be careful because increase pore dimension increase permeability but decrease capillary pressure which manage the working fluid circulation in heat pipe. Therefore the capillary limitation is the main heat transport limitation in wick heat pipe.

Acknowledgements

This article was created within the frame of projects APVV-15-0778 "Limits of radiative and convective cooling through the phase changes of working fluid in loop thermosyphon" and 042ŽU-4/2016 "Cooling on the basis of physical and chemical processes".

Author details

Patrik Nemec

Address all correspondence to: patrik.nemec@fstroj.uniza.sk

University of Zilina, Zilina, Slovakia

References

[1] Peterson GP. An Introduction to Heat Pipes Modelling, Testing, and Applications. 1st ed. New York, USA: John Wiley & Sons. Inc.; 1994

[2] Reay D, Kew P, McGlen R. Heat Pipes–Theory, Design and Applications. 6th ed. Oxford, UK: Elsevier; 2014

[3] Silverstein CC. Design and Technology of Heat Pipes for Cooling and Heat Exchange. 1st ed. Abingdon on Thames, UK: Taylor & Francis; 1992.

[4] Urban F, Kubín M, Kučák L. Experiments on the heat exchangers with the tubes of small diameters. AIP Conference Proceedings. 2014;**1608**:245-248

[5] Ochterbeck JM. Heat pipes, in heat transfer handbook, 1st, July 2003

[6] Swanson LW. Heat Pipe, Heat and Mass Transfer. In: Frank Kreith, editor. Mechanical Engineering Hand-book. Boca Raton, USA: CRC Press LLC; 1999

[7] Chatys R, Malcho M, Orman LJ. Heat transfer enhancement in phase-change heat exchanger. Aviation. 2014;**18**(1):40-43

[8] Zohuri B. Heat Pipe Design and Technology: A Practical Approach. Taylor and Francis Group: Boca Raton; 2011

[9] Peterson GP. An Introduction to Heat Pipes: Modelling, Testing and Applications. Atlanta: Wiley-Interscience; 1994

[10] Silverstein C. Desigh and Technology of Heat Pipes for Cooling and Heat Exchange. Boca Raton: CRC Press; 1992

[11] Faghri A. Heat Pipe Science and Technology. Washington, DC: Taylor & Francis; 1995

[12] Chi SW. Heat Pipe Theory and Practice. Washington, D.C.: Hemisphere Publishing; 1976

[13] Reay D, Kew P. Heat Pipes Theory, Design and Applications. 5th ed. Oxford, UK: Elsevier; 2006

[14] Busse CA. Theory of the ultimate transfer of cylindrical heat pipes. International Journal Heat Mass Transfer. 1973;**16**:169-186

[15] Ivanovskii MN, Sorokin VP, Yagodkin IV. The Physical Properties of Heat Pipes. Oxford: Clarendon Press; 1982

[16] Busse CA, Kemme JE. Dry–out phenomena in gravity–Assist heat pipes with capillary flow. International Journal Heat Mass Transfer. 1980;**23**:643-654

[17] Cotter TP. Heat pipe startup dynamics. Proc. SAE Thermionic Conversion Specialist Conference, Palo Alto, CA, 1967

[18] Marcu BD. On the Operation of Heat Pipes, Report 9895-6001-TU-000. Redondo Beach, CA: TRW; 1965

[19] Dunn PD, Reay DA. Heat Pipes. 3rd ed. New York: Pergamon Press; 1982

[20] Brennan PJ, Kroliczek EJ. Heat Pipe Design Handbook, NASA Contract Report NAS5 – 23406; 1979

[21] Hlavačka V, Polášek F, Štulc P, Zbořil V. Tepelné trubice v elektrotechnice. Praha: STNL - Nakladatelství technické literatúry; 1990

Chapter 8

Analysis of the Influence of Fish Behavior on the Hydrodynamics of Net Cage

Tiao-Jian Xu, Ming-Fu Tang and Guo-Hai Dong

Additional information is available at the end of the chapter

http://dx.doi.org/10.5772/intechopen.71015

Abstract

In net cage hydrodynamic analysis, drag force of net is dependent on the physical dimensions of the net cage, the Solidity ratio, the Reynolds number and the projected area of the net, which is illustrated in numerous previous researches. However, rare studies attempt to investigate the effect of fish behavior. Thus a net-fluid interaction model and a simplified fish model were proposed for analyzing the effects of fish behavior on the net cage. A series of physical model tests were conducted to validate the numerical model, which indicates models can simulate the stocked net cage in the current accurately. The simulation results indicate that circular movement of fish leads to a low pressure zone at the center of net cage, which causes a strong vertical flow along the center line of the net cage. The drag force on the net cage is significantly decreased with the increasing fish stocking density.

Keywords: SST k-omega model, finite element method, net cage, fish behavior

1. Introduction

A slew of researches show that the drag force (F_D) on a net cage is dependent on the physical dimensions of the net cage, the Solidity ratio, the Reynolds number and the projected area of the net. In addition, the fish behavior, such as the fish distribution, the fish swimming velocity and the fish group structure, plays a significant role in the net cage hydrodynamic analysis. Based on the previous site observation in the commercial net cage, Gansel discovered a strong diurnal variation of the fish distribution, indicating that most fishes at night were near the top of net cage, while majorities of fish during daytime were near the bottom of net cage [1]. According to the recent observations on the Atlantic salmon swimming behavior at a commercial farm, Johansson proposed that at the low current velocities (<0.45 BL/s, BL: fish body length), salmon

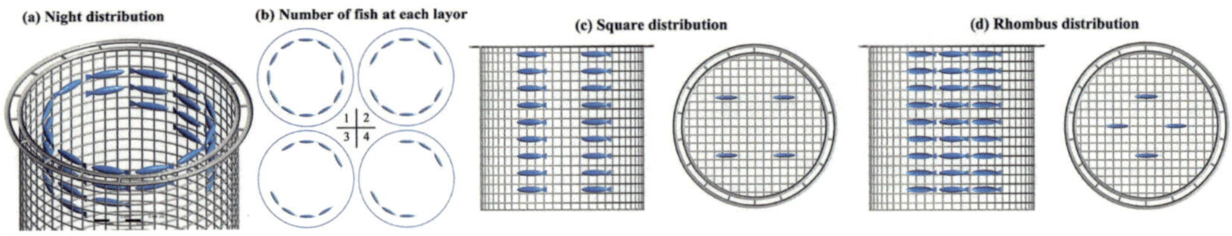

Figure 1. Fish group structure in this chapter (a-d).

group polarized swam in a circular movement and at the high current velocities (>0.90 BL/s), all fishes kept stations at fixed positions swimming against the current [2].

In this chapter, two types of fish distribution patterns were analyzed: night distribution, fish is mainly gathered in the water surface and fish density is decreased with increasing water depth; and day distribution, most fishes perched near to the bottom of net cage and fish density is increased with increasing water depth. And two kinds of fish group structures are considered: circular (**Figure 1**) = polarized swimming in a circular movement; on-current (**Figure 1**) = swimming toward the current without forward movement. The latest progress of net cage numerical simulation method by our research group will be introduced in detail. The contents of the progresses include the following parts: *The introduction of an elaborate net cage model and fish model; The validation of models in steady flow; The numerical simulation of the stocked net cage in steady flow.*

2. Numerical modeling approach

The numerical modeling for analyzing the flow field through the stocked net cage and the deformation of net cage is to combine the *k-ω* Shear Stress Turbulent (SST) model and the large deformation nonlinear structure (LDNS) model. In the combined analysis, the two methods are handled separately.

2.1. Flow around net cage

The *k-ω* SST model, developed by Menter in [3], is applied to simulate the flow filed around the net cage structure.

2.1.1. The k-ω SST model

The governing equations describing the *k-ω* SST turbulence model are as follows:

Continuity equation:

$$\frac{\partial \rho}{\partial t} + \frac{\partial (\rho u_i)}{\partial x_i} = 0 \tag{1}$$

Momentum equation:

$$\frac{\partial(\rho u_i)}{\partial t} + \frac{\partial(\rho u_i u_j)}{\partial x_j} = \frac{-\partial P}{\partial t} + \rho g_i + \frac{\partial}{\partial x_j}(\mu + \mu_t)\left(\frac{\partial u_i}{\partial x_j} + \frac{\partial u_j}{\partial x_i}\right) \quad (2)$$

where ρ is the density of the fluid, u_i and u_j are the time-averaged velocity components for x, y and z, and μ and μ_t are the viscosity of the fluid and the eddy viscosity, respectively. $P=p+(2/3)\rho k$, where p is the time-averaged pressure and P is the transient pressure; k is the turbulent kinetic energy; g_i is the acceleration due to gravity, and $i, j = 1, 2, 3$ (x, y, z).

The k-ω SST model solves two transport equations: the turbulent kinetic energy k and the turbulent dissipation rate ω, as follows:

$$\frac{\partial(\rho k)}{\partial t} + \frac{\partial(\rho u_i k)}{\partial x_i} = \tilde{P}_k - \beta^* \rho k \omega + \frac{\partial}{\partial x_i}\left[(\mu + \sigma_k \mu_t)\frac{\partial k}{\partial x_i}\right] \quad (3)$$

$$\frac{\partial(\rho \omega)}{\partial t} + \frac{\partial(\rho u_i \omega)}{\partial x_i} = \alpha \rho S^2 - \beta \rho \omega^2 + \frac{\partial}{\partial x_i}\left[(\mu + \sigma_\omega \mu_t)\frac{\partial \omega}{\partial x_i}\right] + 2(1 - F_1)\rho \sigma_{\omega 2}\frac{1}{\omega}\frac{\partial k}{\partial x_i}\frac{\partial \omega}{\partial x_i} \quad (4)$$

2.1.2. Boundary conditions

The boundaries of the numerical flume are constituted by a free surface, three wall surfaces, an inlet surface and an outlet surface. The fsi_wall is modeled as a non-slip wall with considering the roughness of the net twine. To consider the roughness of the surface of the actual net twine, the physical roughness height K_s is set as 0.3 mm and the roughness constant C_s is set as 0.5. $y^+ = 1$ is used to determine the thickness of boundary layer.

The governing equations for describing the flow field are solved by a three-dimensional pressure-based Navier-Stokes solver. The SIMPLEC algorithms in [4] are employed to treat the pressure-velocity coupling. The discretization scheme for pressure, momentum, turbulent kinetic energy is carried out using a second order upwind scheme.

2.2. Structural model

The three-dimensional net cage is divided into discrete elements, which the net mesh is modeled as the bar element connected with spherical hinge, as shown in **Figure 2**. The connection force between the net bar and the spherical hinges including two parts: the tangential force and the normal force.

2.2.1. Large deformation nonlinear structure model

The flexible fish net will experience geometric-nonlinear deformation under the action of the hydrodynamic loads, thus the LDNS model (in [5]) is applied here to describe the deformation of flexible fish net. The connection constraints are given as follows:

$$e_j^0 x_i - E_j(X_{c_0} - X_i) = 0(i = 1, 2, 3; j = x, y, z) \quad (5)$$

where a local Cartesian coordinate, x_0-y_0-z_0, is defined at the center of spherical hinge, c_0, and the coordinate system X-Y-Z is the reference coordinate in the LDNS model as shown in

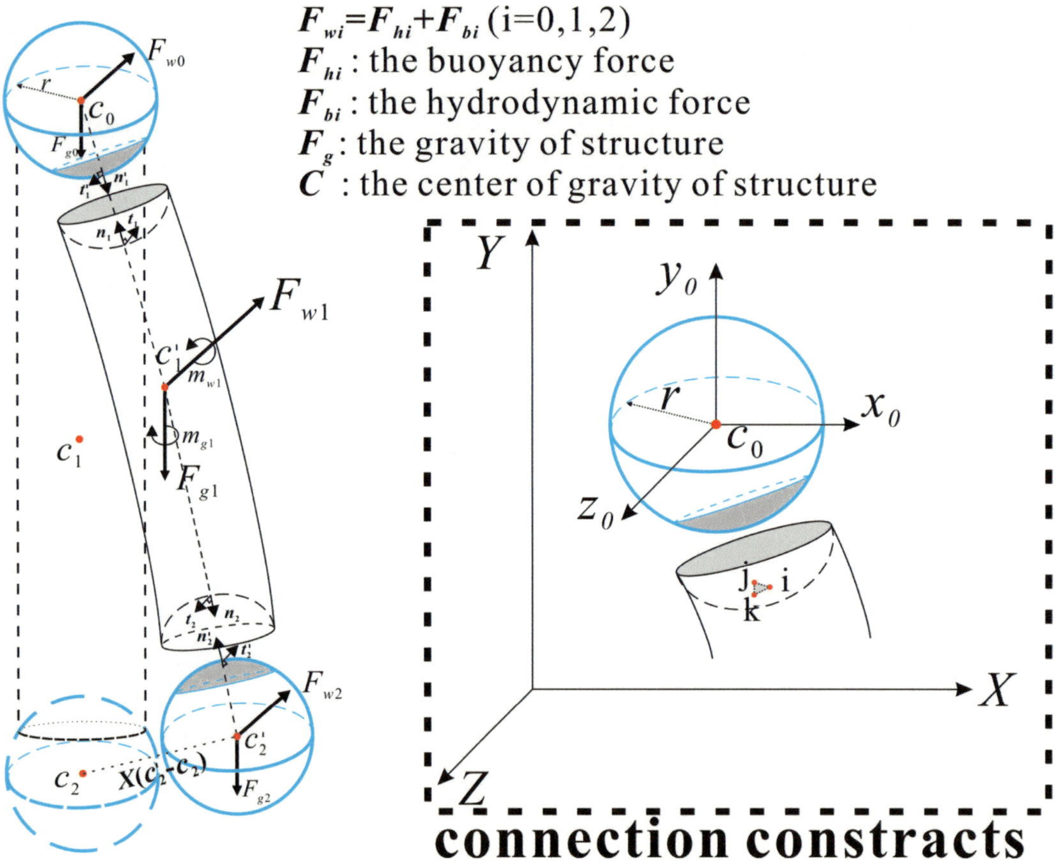

Figure 2. The finite element model of net twine with spherical hinges and mesh bar.

Figure 2; e_j^0, is the unit vector of the local coordinate system C_0-$x_0y_0z_0$ and E_j is the unit vector of the reference coordinate X-Y-Z; the governing equation applied to the LDNS model is given as:

$$[K(x)]x = F_h + F_b + F_g = Q \tag{6}$$

where $K(x)$ is the nonlinear structural stiffness matrix which is related with the unknown displacement vector x.

2.2.2. Load transfer and grid generation

To transfer pressure data from the flow model to the structural model, all nodes on the net twines surface are projected to the fsi_wall element in the flow model according to the mapping rule—projecting each nodes in the target surface normal to the nearest mesh face in the source surface as shown in **Figure 3**. The transferred variable is linearly interpolated on the source face by linear two-dimensional shape function.

http://dx.doi.org/10.5772/intechopen.71015

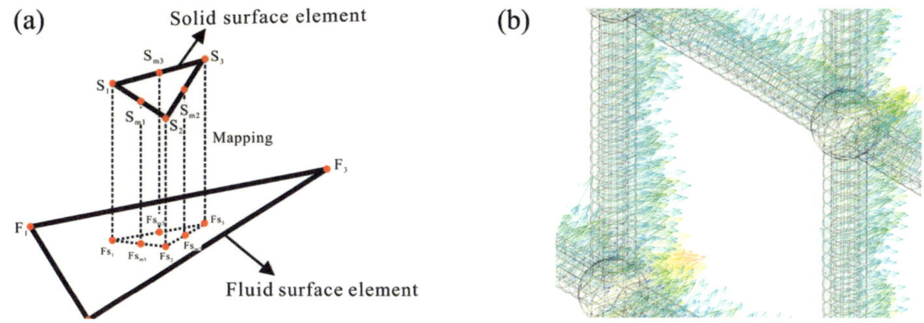

Figure 3. The process of force transfer from fluent solver to structure solver: (a) the sketch of data transfer procedure; (b) the transfer of hydrodynamic force on the net mesh.

2.3. Fish model

As shown in **Figure 4**, the fish model includes the fish body and the fish tail. The fish body suffers the drag force and the fish tail creates the propulsion force. The resulting force of the model includes the propulsion force and the drag force, given as follows:

$$F_x = F_{hx} - F_D cos\beta + F_L sin\beta = \int_{S_t} \tau_{hx} ds - \int_{S_b} \tau_{dx} ds cos\beta = ma_x \tag{7}$$

$$F_y = F_{hy} - F_D sin\beta - F_L cos\beta = \int_{S_t} \tau_{hy} ds - \int_{S_b} \tau_{dy} ds sin\beta = ma_y \tag{8}$$

where the F_D, F_L and F_h are the drag force, lift force and the propulsion force of the model, respectively; τ_d and τ_h are the wall shear stress on the fish body and tail. According to the research (in [6]), the forces on the fish are calculated as follows:

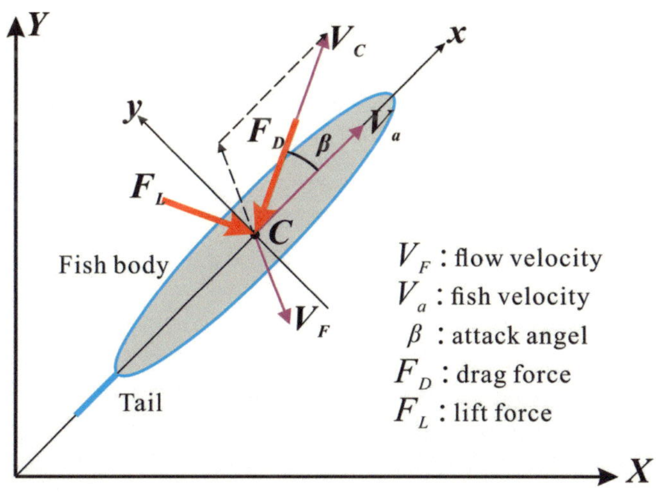

Figure 4. Top view of the tail-actuated fish model.

$$F_D = \frac{1}{2}\rho|V_c|^2 SC_D \; F_L = \frac{1}{2}\rho|V_c|^2 SC_L \tag{9}$$

$$C_D = C'_{D0} + C'_{D1}\alpha_0^2 \; C_L = C'_{L\beta}\alpha_0 = 0.935\alpha_0; \tag{10}$$

where S is the surface area for the fish body, V_c is the relative velocity of fish, β is the attack angle of fish body, C_D and C_L are the drag coefficient and the lift coefficient, C'_{D0} and C'_{D1} are empirical constants equal to 0.1936 and 0.1412 rad^{-2} according to [6], α_0 is the swinging angle of fish tail.

2.4. Grid generation

An example of computational grids for a plane net is shown in **Figure 5(a)**. Tetrahedral grids exist in the majority computational area, and pentahedral prism grids are adopted to refine the meshes near the fluid-solid boundary. Ten boundary mesh layers are adopted to generate boundary grids around fish model as shown in **Figure 5(b)**. Bar elements are divided by hexahedral grids with uniform grid and spherical hinges are divided using tetrahedral grids as shown in **Figure 5(c)**.

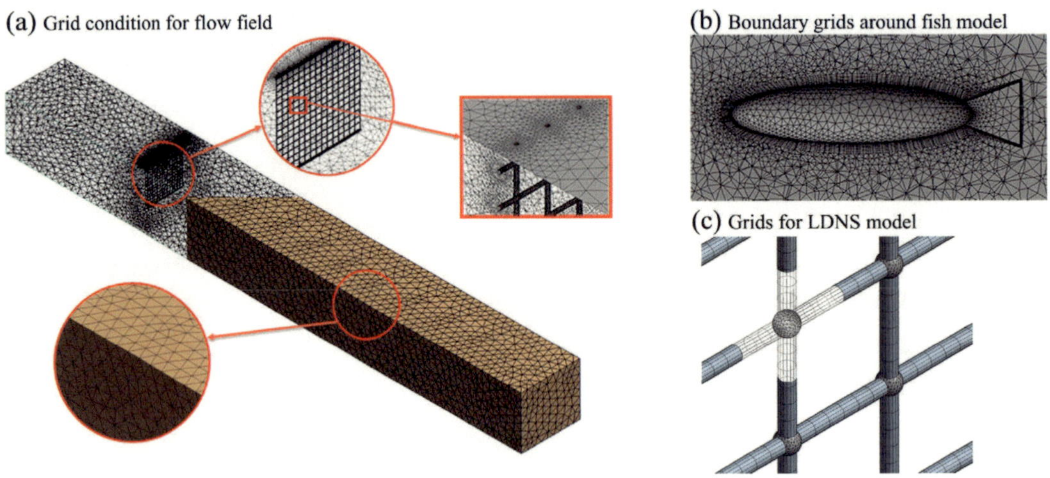

Figure 5. Grid conditions used in this chapter.

3. Experimental validations of model

For fish model validation, a rigid fish model with different behaviors is simulated and the results are compared with the formula proposed in [6]. To validate the net cage model, a series of net cage experiment are conducted a flume at the State Key Laboratory of Coastal and Offshore Engineering, Dalian University of Technology, Dalian, China. The flume is 22 m long, 0.45 m wide and 0.4 m water depth in the tests.

3.1. Validation of the fish model

A single rigid fish model is modeled in the flow as shown in **Figure 6**. The size and weight of fish are same as small fish model (in **Table 1**). The specified wall shear stress in Eqs. (7) and (8), is used to simulate different fish behaviors. There are three operating variables: fish yaw angle, flow velocity and fish swimming velocity, used to investigate the effects of the fish behavior on the flow field and the net deformation. Thus three experiment groups are produced to validate fish model (referring to **Table 2**). **Figure 7** shows that simulation is in good agreement with Eq. (9).

3.2. Validation of the net cage

To validate the deformation of net cage in the steady flow, a net cage model is tested in the flume. The diameter of net cage is 0.254 m and the net mesh is square with 20.0 mm mesh size and 1.2 mm twine diameter. The top of the net is mounted on a top steel ring shown in **Figure 8**.

Figure 9 indicates that the cage deformation simulation is close to the experiment. **Figure 10** indicates the drag force on the net cage from simulation is in good agreement with experiment and the relative error is less than 6.82%.

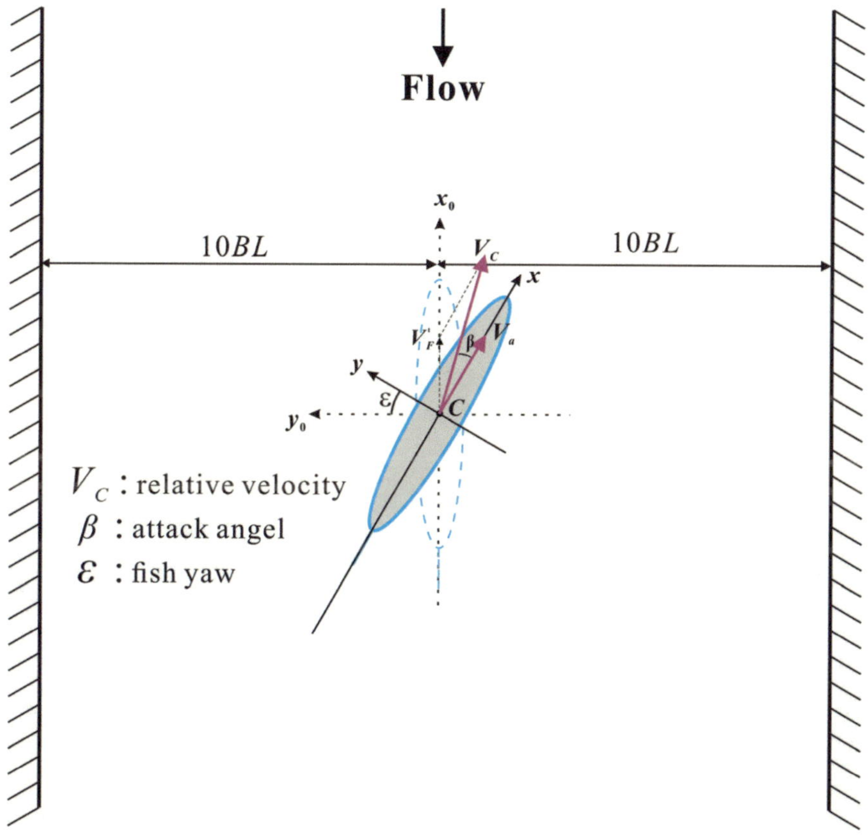

Figure 6. Sketch of the force on single fish model with different yaw in current.

Items	Small fish		Large fish	
	Real fish	Fish model	Real fish	Fish model
Fish length (cm)	40	6.7	50	8.5
Fish surface area (cm^2)	–	1.235	–	2.112

Table 1. The characteristic parameters of fish models applied in the simulation.

No.	V_F (cm/s)	V_a (cm/s)	V_C (cm/s)	Yaw (rad)	β (rad)
1	0	2.0, 3.0, 4.0, 6.0, 8.0, 10.0	2.0, 3.0, 4.0, 6.0, 8.0, 10.0	0	0
2	3.0, 6.0, 6.2	0	3.0, 6.0, 6.2	0	0
3	3.0	6.0	–	0-π	–

***Note*:** V_C and β are determined by fish yaw, flow velocity (V_F) and fish velocity (V_a).

Table 2. Experiment groups of fish model.

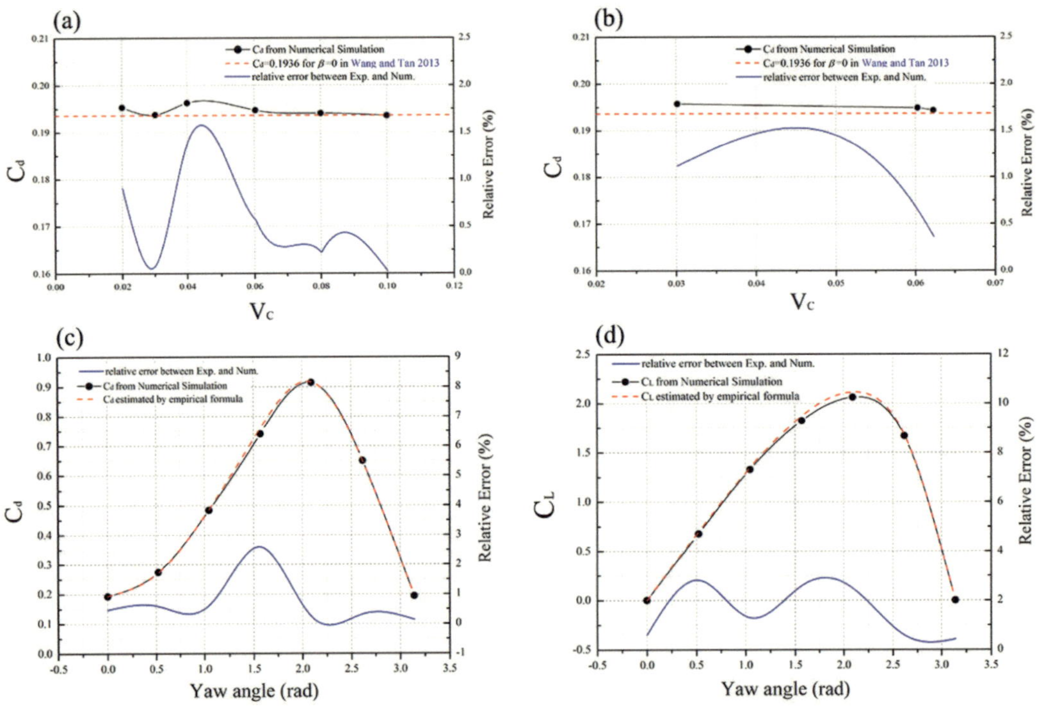

Figure 7. Comparison between simulation and empirical formula: (a) the drag coefficient with different fish velocities in still water; (b) the drag coefficient of fish in different flow velocities; (c) and (d) are the force coefficient of fish for different yaw angles.

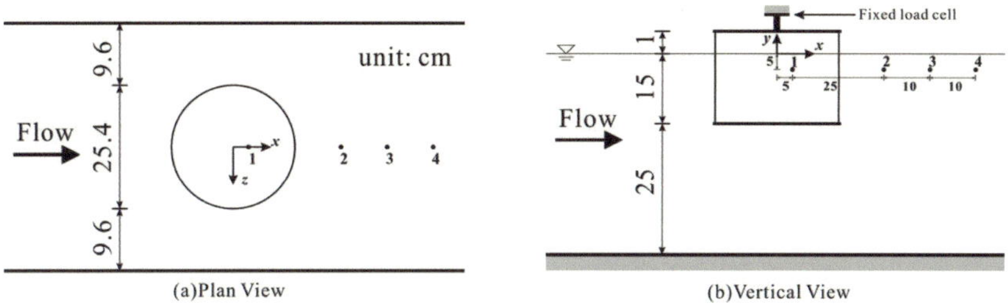

Figure 8. Physical model of net cage and general setting of the measurement points (a, b).

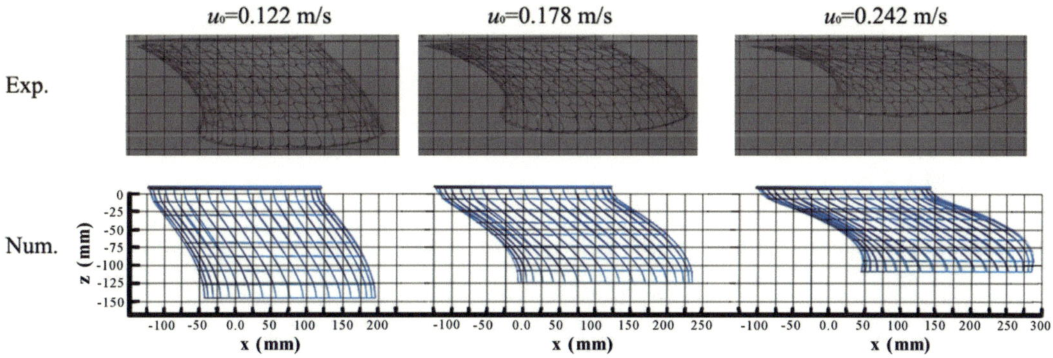

Figure 9. Comparison of the deformation of the net cage between the simulation and experiment.

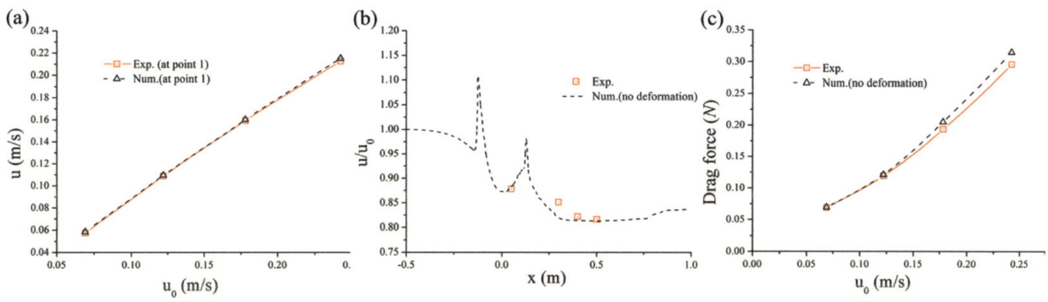

Figure 10. Comparisons of simulation and experiment: (a) flow velocities at point 1 for different incoming velocities; (b) dimensionless flow velocities inside and around net cage with 0.242 m/s incoming velocity; (c) drag forces on net cage for different flow velocities.

4. The numerical simulation of the stocked net cage in steady flow

To investigate the effect of fish behavior on the net cage, a 2.5 m long, 2.5 m wide and 0.7 m depth numerical water flume is adopted here, and the stocked net cage model with 0.35 m outer diameter and 0.35 m height is located at the center of the flume. Sinkers with 1 N weight are applied to maintain the shape of net cage. The net is mounted as square meshes, in which

No.	*SD* (kg/m^3)	Fish model	V_F (BL/s)	Distribution	GS	V_a (BL/s)	A_a (BL/s^2)
1	8	Small	0	Night/day	Circle	0.45 & 0.90 0.45	0 0.45
2	0 8	– Small	0.45	– Night/day	– Circle	– 0.90	– 0
3	0 8	– Small	0.93	– Square/rhombus	– On-current	– 0.93	– 0
4	8 16	Large	0.45	Night/day	Circle	0.90	0
5	24, 32, 40	Large	0.45	Uniform	Circle	0.90	0

BL is the body length of fish; *SD* is stocking density; *GS* is group structure.

Table 3. Experiment groups of fish with different stocking densities and fish distributions.

the twine diameter is 2.6 mm and the mesh bar length is 20 mm. The detailed experiment setup is shown in **Table 3**.

4.1. Circular movement of fish in the still water

The effect of fish swimming on the flow pattern around the net cage in the still water is analyzed, in EG (experiment group) 1 in **Table 3**. **Figure 11** shows the flow pattern in the net cage for different fish distributions with 0.45 BL/s swimming speed. According to Newton's second law, the centripetal force acting on fish need be balanced by the hydrodynamic force which pushes water away from fish. Thus the flow velocity near the inner boundary of fish is greater than that of the outer boundary in the horizontal plane, leading to a low pressure area in the center of the rotational movement of the fish. For the day distribution, water is pulled from above and below the fish swimming depth; for the night distribution, water is only pulled from below the fish swimming depth.

4.2. Effect of fish distribution on flow field and drag force

The effects of fish distribution on the flow field around the net cage and the drag force of the net cage are analyzed in EG 2 and 3. **Figure 12** shows the flow field around the net cage for the low current velocity case (EG 3) and the fish motion in high current case (EG 4) has little influence on

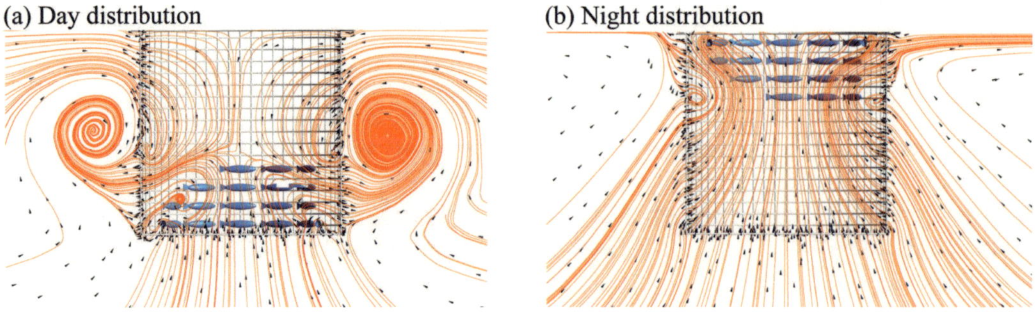

Figure 11. The flow around fish cage for different distributions in still water with 0.45 BL/s fish speed (a, b).

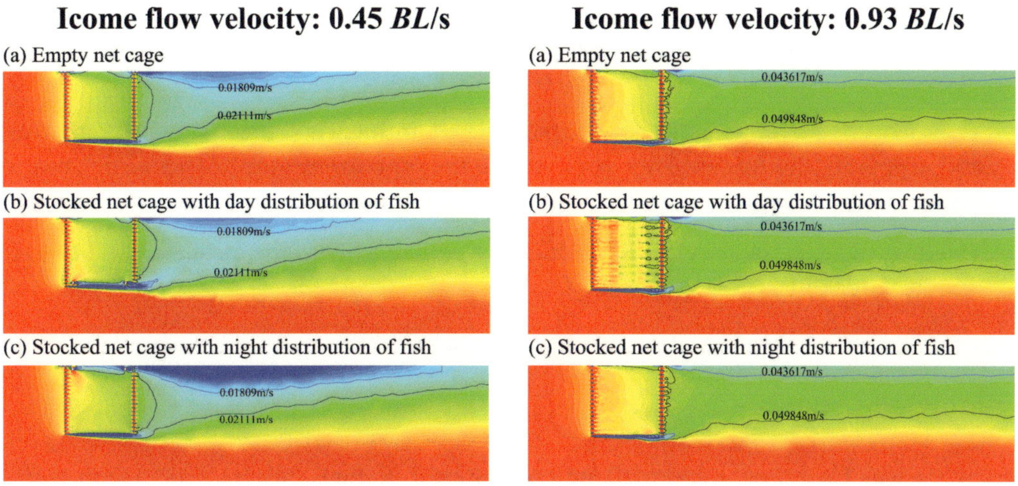

Figure 12. The flow field around the empty net cage and the stocked net cage in low current case (V0 = 0.45 BL/s) and high current case (V0 = 0.93 BL/s) (a-c).

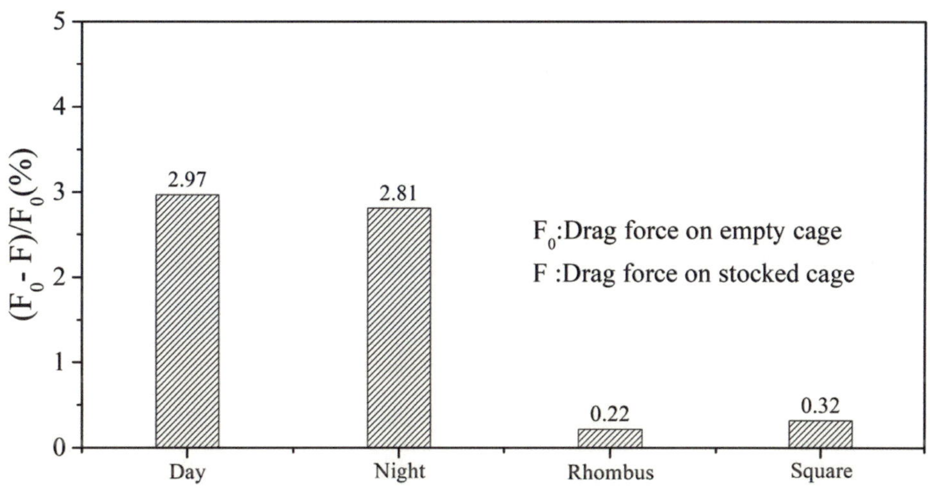

Figure 13. Normalized drag force on the stocked cage for different fish distribution.

the downstream. **Figure 13** shows the circular movement of fish has larger influence on the drag force acting on the net cage and the influence of on-current movement is little.

5. Conclusions

A net-fluid interaction model and a simplified fish model are proposed for analyzing the effects of fish behavior on the flow field around the net cage and the deformation of the net cage. And the following conclusions can be drawn from the case study:

1. The fish circular movement around the net cage can produce a low pressure zone at the center of the net cage, causing a vertical water exchange along the center line of the net cage.
2. The circular movement of fish has significant influence on the downstream wake, and especially the low-velocity zone. While the on-current movements of fishes affect little.
3. The drag force on the net cage is significantly decreased with the increasing fish stocking density. Little differences in the fish distribution are observed.

Acknowledgements

This work was financially supported by the National Natural Science Foundation (NSFC) Projects No. 51239002, 51409037, 51579037, and 51221961, China Postdoctoral Science Foundation (No. 2014M560211 and No. 2015T80254), the Fundamental Research Funds for the Central Universities No. DUT16RC(4)25 and Cultivation plan for young agriculture science and technology innovation talents of Liaoning province (No. 2014008).

Author details

Tiao-Jian Xu*, Ming-Fu Tang and Guo-Hai Dong

*Address all correspondence to: tjxu@dlut.edu.cn

Dalian University of Technology, China

References

[1] Gansel LC, Rackebrandt S, Oppedal F, McClimans TA. Flow fields inside stocked fish cages and the near environment. In: 30-th International Conference on Ocean, Offshore and Arctic Engineering. (OMAE 2011-50205); 2011

[2] Johansson D, Laursen F, FernÖ A, Fosseidengen JE, Klebert P, Stien LH, Vågseth T, Oppedal F. The interaction between water currents and salmon swimming behavior in sea cages. PLoS One. 2014;**9**(5):e97635

[3] Menter FR. Two-equation eddy-viscosity turbulence models for engineering applications. AIAA-Journal. 1994;**32**(8):269-289

[4] Vandoormaal JP, Raithby GD. Enhancements of the SIMPLE method for predicting incompressible fluid flows. Numerical Heat Transfer. 1984;**7**:147-163

[5] Bathe KJ, Ramm E, Wilson EL. Finite element formulations for large deformation dynamic analysis. International Journal for Numerical Methods in Engineering. 1975;**9**:353-386

[6] Wang JX, Tan XB. A dynamic model for tail-actuated robotic fish with drag coefficient adaptation. Mechatronics. 2013;**23**:659-668

Section 4

Quantum Gravity

Chapter 9

Beyond Einstein: A Polynomial Affine Model of Gravity

Oscar Castillo-Felisola

Additional information is available at the end of the chapter

http://dx.doi.org/10.5772/intechopen.70951

Abstract

We show that the effective field equations for a recently formulated polynomial affine model of gravity, in the sector of a torsion-free connection, accept general Einstein manifolds—with or without cosmological constant—as solutions. Moreover, the effective field equations are partially those obtained from a gravitational Yang-Mills theory known as the Stephenson-Kilmister-Yang (SKY) theory. Additionally, we find a generalisation of a minimally coupled massless scalar field in general relativity within a "minimally" coupled scalar field in this affine model. Finally, we present the road map to finding general solutions to the effective field equations with either isotropic or cosmologic (i.e., homogeneous and isotropic) symmetry.

Keywords: polynomial affine gravity, torsion, generalised gravity

1. Introduction

During the last century (approximately), we reached a high level of understanding of the four fundamental interactions, that is, electromagnetic, weak, strong and gravitational. However, our understanding splits into two streams: the first describes three kinds of interactions and includes them into a single model, called standard model of particle physics, while the second covers only the gravitational interactions.

The interactions within the standard model of particles are described by connection fields, modelled by gauge theories, and their quantisation procedure is successfully applied. On the other hand, the gravitational interaction, as formulated by Einstein [1] and Hilbert [2], is described by the metric field, whose model does not fit into the category of gauge theory, and its quantisation procedure is not yet well defined [3–9] (see [10] for a historical review).

The above suggests that some of the theoretical problems encountered when trying to quantise the gravitational interactions[1] are due to fact that it is formulated as a field theory for the metric and not as a theory for a connection. Therefore, there have been several attempts of describing the gravitational interaction from an affine view point, by using only the connection as fundamental field of the model [11, 12], but these descriptions were not very successful. In addition, Cartan's proposal of considering the same field equations (or action) than Einstein-Hilbert, but with more general connections, see [13–16], was left aside because the field equations, in pure gravity, impose the vanishing torsion.

After the works by Kibble [17] and Sciama [18], it was understsood that once gravity couples to matter, a nontrivial torsion compatible with the setup could exist and the search of viable affine [19–30] (and metric-affine [31]) models of gravity becomes relevant again.

Before continuing our arguments for the *necessity* of considering affine generalisations of general relativity, we shall briefly remind some basic concepts in geometry. In the antiquity, the plane geometry was built essentially with the aid of a (straight) rule and a compass. Counterintuitively, the compass was used to measure distances, while the rule was used to define parallelism. In modern differential geometry language, the object that allows us to measure distances is the metric, ($g_{\mu\nu}$), while the one associated with the concept of parallelism is the connection ($\Gamma^{\lambda}{}_{\mu\nu}$).

Although the concepts of distance and parallelism are independent, there exists a (unique) particular case in which the concepts relate with each other, and thus one needs just of the metric: while the connection is a potential for the metric. This particular case is known as Riemannian geometry, and general relativity stands on such particular construction.

The geometries with general connection are called with the adjective affine and are characterised by its curvature ($\mathcal{R}_{\mu\nu}{}^{\lambda}{}_{\rho}$), torsion ($\mathcal{T}^{\lambda}{}_{\mu\nu} = \Gamma^{\lambda}{}_{\mu\nu} - \Gamma^{\lambda}{}_{\nu\mu}$), and the metricity condition ($\mathcal{Q}_{\lambda\mu\nu} = \nabla^{\Gamma}_{\lambda} g_{\mu\nu}$) [31–33]. Notice that in Riemannian geometries, either $\mathcal{T}^{\lambda}{}_{\mu\nu}$ or $\mathcal{Q}_{\lambda\mu\nu}$ vanishes, and the only quantity that characterises the manifolds is the curvature.

An argument to consider affine geometries, to describe gravitational interactions, is that it introduces new degrees of freedom into the model (other than a spin-two field), which might be interpreted as matter—from the view point of general relativity—contributing to the dark sector of the universe [34, 35], inflation [36], exotic cosmologies [37], and diverse particle physics effects [38–46].

In the following sections, we will analyse a recently proposed model called *polynomial affine gravity* [29, 30], which is built up with an affine connection as sole field, and under the premise of preserving the whole group of diffeomorphisms.

2. Polynomial affine gravity: The model

The connection is the field that allows us to define the notion of parallelism as follows. Given a connection $\hat{\Gamma}^{\mu}_{\rho\sigma}$, one defines a *covariant* derivative, ($\nabla^{\hat{\Gamma}}$), such that, if the directional derivative

[1] The issue is not only related to quantization, but also to generality. Mathematically, the only ingredient needed to define the curvature is the connection. Thus, considering connections defined by a more fundamental field is not the most general set up.

http://dx.doi.org/10.5772/intechopen.70951

of a geometrical object, $\mathcal{V}$, along a vector (X) vanishes, one says that the object is parallel transported along (the integral curve defined by) the vector

$$\nabla_X^{\widehat{\Gamma}}\mathcal{V} = 0.$$

The affine connection accepts a decomposition on irreducible components as

$$\widehat{\Gamma}^{\mu}_{\rho\sigma} = \widehat{\Gamma}^{\mu}_{(\rho\sigma)} + \widehat{\Gamma}^{\mu}_{[\rho\sigma]} = \Gamma^{\mu}{}_{\rho\sigma} + \varepsilon_{\rho\sigma\lambda\kappa}T^{\mu,\lambda\kappa} + A_{[\rho}\delta^{\mu}_{\nu]}, \tag{1}$$

where $\Gamma^{\mu}{}_{\rho\sigma} = \widehat{\Gamma}^{\mu}_{(\rho\sigma)}$ is symmetric in the lower indices, A_μ is a vector field corresponding to the trace of torsion and $T^{\mu,\lambda\kappa}$ is a Curtright-like field [47],[2] satisfying $T^{\kappa,\mu\nu} = -T^{\kappa,\nu\mu}$ and $\varepsilon_{\lambda\kappa\mu\nu}T^{\kappa,\mu\nu} = 0$.[3]

Using the above decomposition, we need to build the most general action preserving diffeomorphisms. In order to guarantee the correct transformation of the Lagrangian density, the geometrical objects used to write down the action are a Curtright ($T^{\mu,\nu\lambda}$), a vector (A_μ), the covariant derivative defined with the Levi-Civita connection (∇_μ), both Levi-Civita tensors ($\varepsilon_{\mu\nu\lambda\rho}$ and $\varepsilon^{\mu\nu\lambda\rho}$) and the Riemannian curvature ($R_{\mu\nu}{}^{\lambda}{}_{\rho}$). Since the Riemannian curvature is defined as the commutator of the covariant derivative, it is not an independent field, so it will be left out of the analysis, and only five ingredients remain. In [30], a method of *dimensional analysis* was introduced to ensure that all possible terms were taken into account, and the general action—up to boundary and topological terms—is

$$\begin{aligned} S[\Gamma,T,A] = \int \mathrm{d}^4x \Big[& B_1 R_{\mu\nu}{}^{\mu}{}_{\rho} T^{\nu,\alpha\beta} T^{\rho,\gamma\delta}\epsilon_{\alpha\beta\gamma\delta} + B_2 R_{\mu\nu}{}^{\sigma}{}_{\rho} T^{\beta,\mu\nu} T^{\rho,\gamma\delta}\epsilon_{\sigma\beta\gamma\delta} + B_3 R_{\mu\nu}{}^{\mu}{}_{\rho} T^{\nu,\rho\sigma} A_\sigma + B_4 R_{\mu\nu}{}^{\sigma}{}_{\rho} T^{\rho,\mu\nu} A_\sigma \\ & + B_5 R_{\mu\nu}{}^{\rho}{}_{\rho} T^{\sigma,\mu\nu} A_\sigma + C_1 R_{\mu\rho}{}^{\mu}{}_{\nu} \nabla_\sigma T^{\nu,\rho\sigma} + C_2 R_{\mu\nu}{}^{\rho}{}_{\rho} \nabla_\sigma T^{\sigma,\mu\nu} + D_1 T^{\alpha,\mu\nu} T^{\beta,\rho\sigma} \nabla_\gamma T^{(\lambda,\kappa)\gamma} \epsilon_{\beta\mu\nu\lambda}\epsilon_{\alpha\rho\sigma\kappa} \\ & + D_2 T^{\alpha,\mu\nu} T^{\lambda,\beta\gamma} \nabla_\lambda T^{\delta,\rho\sigma} \epsilon_{\alpha\beta\gamma\delta}\epsilon_{\mu\nu\rho\sigma} + D_3 T^{\mu,\alpha\beta} T^{\lambda,\nu\gamma} \nabla_\lambda T^{\delta,\rho\sigma} \epsilon_{\alpha\beta\gamma\delta}\epsilon_{\mu\nu\rho\sigma} + D_4 T^{\lambda,\mu\nu} T^{\kappa,\rho\sigma} \nabla_{(\lambda} A_{\kappa)} \epsilon_{\mu\nu\rho\sigma} \\ & + D_5 T^{\lambda,\mu\nu} \nabla_{[\lambda} T^{\kappa,\rho\sigma} A_{\kappa]} \epsilon_{\mu\nu\rho\sigma} + D_6 T^{\lambda,\mu\nu} A_\nu \nabla_{(\lambda} A_{\mu)} + D_7 T^{\lambda,\mu\nu} A_\lambda \nabla_{[\mu} A_{\nu]} \\ & + E_1 \nabla_{(\rho} T^{\rho,\mu\nu} \nabla_{\sigma)} T^{\sigma,\lambda\kappa} \epsilon_{\mu\nu\lambda\kappa} + E_2 \nabla_{(\lambda} T^{\lambda,\mu\nu} \nabla_{\mu)} A_\nu + T^{\alpha,\beta\gamma} T^{\delta,\eta\kappa} T^{\lambda,\mu\nu} T^{\rho,\sigma\tau} (F_1 \epsilon_{\beta\gamma\eta\kappa}\epsilon_{\alpha\rho\mu\nu}\epsilon_{\delta\lambda\sigma\tau} \\ & + F_2 \epsilon_{\beta\lambda\eta\kappa}\epsilon_{\gamma\rho\mu\nu}\epsilon_{\alpha\delta\sigma\tau}) + F_3 T^{\rho,\alpha\beta} T^{\gamma,\mu\nu} T^{\lambda,\sigma\tau} A_\tau \epsilon_{\alpha\beta\gamma\lambda}\epsilon_{\mu\nu\rho\sigma} + F_4 T^{\eta,\alpha\beta} T^{\kappa,\gamma\delta} A_\eta A_\kappa \epsilon_{\alpha\beta\gamma\delta} \Big]. \end{aligned} \tag{2}$$

Despite the complex structure of the action, it shows very interesting features: (a) The structure is rigid (it does not accept extra terms), which forbids the appearance of counter-terms, if one would like to quantise the model; (b) all coupling constants are dimensionless, which might be a hint of conformal invariance of the model; (c) the action turns out to be power-counting renormalisable, which does not guarantee renormalisability, but is a nice feature; and (d) the structure of the model yields no three-point graviton vertices, which might allow to overcome the *no-go* theorems found in Refs. [48, 49].

---

[2] Notice that the Curtright-like field is defined as the quasi-Hodge dual of the traceless part of the torsion.

[3] Since no metric is present, the epsilon symbols are not related by raising (lowering) their indices, but instead we demand that $\varepsilon^{\delta\eta\lambda\kappa}\varepsilon_{\mu\nu\rho\sigma} = 4!\,\delta^{\delta}{}_{[\mu}\delta^{\eta}{}_{\nu}\delta^{\lambda}{}_{\rho}\delta^{\kappa}{}_{\sigma]}$.

3. Limit of vanishing torsion

We now want to restrict ourselves to the limit of vanishing torsion, which simplifies the comparison between our model and general relativity. The vanishing torsion limit—equivalent to take $T^{\lambda,\,\mu\nu} \to 0$ and $A_\mu \to 0$—cannot be taken at the action level, but in the field equations, and the limit is a consistent truncation of the whole field equations [30].

The only nontrivial field equation after the limit will be the one for the Curtright-like field, $T^{\nu,\,\mu\rho}$:

$$\nabla_{[\rho}R_{\mu]\nu} + \kappa\nabla_\nu R_{\mu\rho}{}^\lambda{}_\lambda = 0, \tag{3}$$

with κ a constant related with the original couplings of the model. These field equations are simpler if one restricts to connections compatible with a volume form, also known as equi-affine [32, 50, 51], which assures that the Ricci tensor of the connection is symmetric and the contraction of the last indices vanishes; thus the equation is

$$\nabla_{[\rho}R_{\mu]\nu} = 0. \tag{4}$$

Eq. (4) is a generalisation of Einstein's field equation in vacuum. This can be seem as follows: all Einstein manifolds posses Ricci tensor proportional to the metric, $R_{\mu\nu} \propto g_{\mu\nu}$; the metricity condition thus ensures that every vacuum solution to the Einstein's equations solves the (simplified) field equations of our model.

Moreover, Eq. (4) is related through the second Bianchi identity to the harmonic curvature condition [52]:

$$\nabla_\lambda R_{\mu\nu}{}^\lambda{}_\rho = 0. \tag{5}$$

Eqs. (4) and (5) accept a geometrical interpretation equivalent to that of the field equations of a pure Yang-Mills theory, which in the language of differential forms are

$$\mathcal{D}\boldsymbol{F} = 0, \quad \mathcal{D}\star\boldsymbol{F} = 0, \tag{6}$$

where $\boldsymbol{F} = \mathcal{D}\boldsymbol{A}$ is the field strength two-form (the curvature two-form of the connection in the principal bundle; see, for example, [53, 54]), and the operator $\star\star$ denotes the Hodge star. Now, these Yang-Mills field equations are obtained from the variation of the action functional:

$$S_{\mathrm{YM}} = \int \mathrm{Tr}(\boldsymbol{F}\star\boldsymbol{F}), \tag{7}$$

and the Jacobi identity for the covariant derivative.

Similarly, Eq. (4) (equivalently Eq. (5)) can be obtained from an effective gravitational Yang-Mills functional action [55–57]:

$$S_{\text{YM}} = \int \text{Tr}(\mathcal{R}\star\mathcal{R}) = \int \left(\mathcal{R}^{a}{}_{b}\star\mathcal{R}^{b}{}_{a}\right), \tag{8}$$

where $\mathcal{R} \in \Omega^2(\mathcal{M}, T^*\mathcal{M} \otimes T\mathcal{M})$ is the curvature two-form, the operator $\star$ denotes the Hodge star and the trace is taken on the bundle indices (see [53]).

The gravitational model described by the action in Eq. (8) is called Stephenson-Kilmister-Yang (or SKY for short), and its physical interpretation relies—as in general relativity—in the fact that the metric is the fundamental field for describing the gravitational interaction. In that case, the field equations are third-order partial differential equations, and there are several undesirable behaviours due to this characteristic of the equations. However, in our model the field mediating the gravitational interaction is the connection, and therefore, the Eqs. (4) and (5) are second-order field equations for the components of the connection.

It is worth noticing that, according to the arguments in [48, 49], the SKY theory is not renormalisable. However, it is possible that the polynomial affine gravity could be renormalisable, in the sense that SKY is an effective description for the torsionless limit of polynomial affine gravity.

4. Polynomial affine gravity coupled to a scalar field

In the standard formulation of physical theories (even in flat spacetimes), the metric is a required ingredient. The metric and its inverse define a homomorphism between the tangent and cotangent bundles, allowing to build the kinetic energy term in the action.

$$\partial_\mu\phi\partial^\mu\phi = g^{\mu\nu}\partial_\mu\phi\partial_\nu\phi.$$

Therefore, the inclusion of matter within models with no necessity of a metric is a nontrivial task.

Inspired in the method of *dimensional analysis* introduced in Ref. [30], we attempt to couple a scalar field to the polynomial affine gravity by defining the most general, symmetric $\binom{2}{0}$-tensor density, $\mathfrak{g}^{\mu\nu}$, built with the available fields, and use it to build Lagrangian densities for the matter content. It can be shown that such a density is given by

$$\mathfrak{g}^{\mu\nu} = \alpha\, \nabla_\lambda T^{\mu,\nu\lambda} + \beta\, A_\lambda T^{\mu,\nu\lambda} + \gamma\, \varepsilon_{\lambda\kappa\rho\sigma} T^{\mu,\lambda\kappa} T^{\nu,\rho\sigma}, \tag{9}$$

with α, β and γ arbitrary coefficients.

Therefore, the action defined by the "kinetic term" is

$$S_\phi = -\int \mathrm{d}^4x \left(\alpha\, \nabla_\lambda T^{\mu,\nu\lambda} + \beta\, A_\lambda T^{\mu,\nu\lambda} + \gamma\, \varepsilon_{\lambda\kappa\rho\sigma} T^{\mu,\lambda\kappa} T^{\nu,\rho\sigma}\right)\partial_\mu\phi\partial_\nu\phi. \tag{10}$$

Remarkably, it induces a nontrivial contribution to the field equations once we restrict to the torsionless sector. The field nontrivial equations, when the scalar field is turned on, is[4]

$$\nabla_{[\sigma}R_{\rho]\mu}{}^{\mu}{}_{\nu} - C_2\nabla_{\nu}R_{\rho\sigma}{}^{\mu}{}_{\mu} - \alpha\nabla_{[\sigma}\left(\partial_{\rho]}\phi\partial_{\nu}\phi\right) = 0,$$

which for equi-affine connections simplifies to

$$\nabla_{[\sigma}R_{\rho]\nu} - \alpha\nabla_{[\sigma}\left(\partial_{\rho]}\phi\partial_{\nu}\phi\right) = 0. \tag{11}$$

Eq. (11) can be integrated once, and the solution takes the familiar form:

$$R_{\mu\nu} - \alpha\partial_{\mu}\phi\partial_{\nu}\phi = \Lambda g_{\mu\nu},$$

where the integration (covariantly) constant tensor, which is invertible and symmetric, has been suggestively denoted by $\Lambda g_{\mu\nu}$. The above equation can be written in the more conventional form:

$$R_{\mu\nu} - \frac{1}{2}g_{\mu\nu}R + \Lambda g_{\mu\nu} = \alpha\left(\partial_{\mu}\phi\partial_{\nu}\phi - \frac{1}{2}g_{\mu\nu}(\partial\phi)^2\right). \tag{12}$$

Thus, we have been able to recover a set of equations, similar to those of general relativity coupled with a free, massless scalar field, but with an arbitrary rank-two, symmetric, covariantly constant tensor playing the role of a metric.

Notice that, from the action in Eq. (10), it is not possible to obtain the field equation for the scalar ϕ, when the vanishing torsion limit is taken. Nonetheless, the second Bianchi identity in Eq. (4) imposes

$$\nabla^{\mu}\partial_{\mu}\phi = 0. \tag{13}$$

This condition is, in the sense argued in Ref. [58], the equation of motion for the scalar field.

5. Finding symmetric ansätze

The usual procedure for solving Einstein's equation is to propose an ansatz for the metric. That ansatz must be compatible with the symmetries we would like to respect in the problem. The formal study of the symmetries of the fields is accomplished via the Lie derivative (for reviews, see [54, 59–61]). Below, we use the Lie derivative for obtaining ansätze for either the metric or the connection.

The form of the Lie derivative for tensors is well known, but the Lie derivative for a connection is not. Thus, for the sake of completeness, we remind the readers that

---

[4]We have fixed the coefficient $C_1 = 1$.

http://dx.doi.org/10.5772/intechopen.70951

$$\mathcal{L}_\xi \Gamma^a{}_{bc} = \xi^m \partial_m \Gamma^a{}_{bc} - \Gamma^m{}_{bc} \partial_m \xi^a + \Gamma^a{}_{mc} \partial_b \xi^m + \Gamma^a{}_{bm} \partial_c \xi^m + \frac{\partial^2 \xi^a}{\partial x^b \partial x^c}, \tag{14}$$

where ξ is the vector defining the symmetry flow.

We shall restrict ourselves to the isotropic (spherically symmetric) and homogeneous and isotropic (cosmological symmetry). In the tedious task of calculating the Lie derivative of different objects, we have used the mathematical software SAGE together with its differential geometry package SageManifolds [62, 63].

5.1. Isotropic ansätze

The two-dimensional sphere, S^2, is a maximally symmetric space whose Killing vectors generate an $SO(3)$ symmetry group. The se vectors can be expressed in spherical coordinates as

$$\begin{aligned} J_1 &= \begin{pmatrix} 0 & 0 & -\cos(\varphi) & \cot(\theta)\sin(\varphi) \end{pmatrix}, \\ J_2 &= \begin{pmatrix} 0 & 0 & \sin(\varphi) & \cos(\varphi)\cot(\theta) \end{pmatrix}, \\ J_3 &= \begin{pmatrix} 0 & 0 & 0 & 1 \end{pmatrix}. \end{aligned} \tag{15}$$

5.1.1. Isotropic (covariant) two tensor

Let us start by finding the most general isotropic, four-dimensional, covariant rank-two tensor. We shall obtain a generalisation of the famous ansatz for the Schwarzschild metric.

We start from a general rank-two tensor, that is, the 16 components of the tensor depend on all the coordinates. Then, the Lie derivative of the metric along the vector J_3 in Eq. (15) yields

$$\pounds_{J_3} T_{\mu\nu} = \frac{\partial T_{\mu\nu}}{\partial \varphi}, \tag{16}$$

which vanishes only if none of the components of the metric depend on the φ coordinate.

The Lie derivative along the other two generators of the angular momentum yields a nontrivial set of differential equations (not shown here) whose solution fixes a tensor of the form:

$$T = \begin{pmatrix} T_{00}(t,r) & T_{01}(t,r) & 0 & 0 \\ T_{10}(t,r) & T_{11}(t,r) & 0 & 0 \\ 0 & 0 & T_{22}(t,r) & H(t,r)\sin(\theta) \\ 0 & 0 & -H(t,r)\sin(\theta) & T_{22}(t,r)\sin^2(\theta) \end{pmatrix}. \tag{17}$$

This result was found by Papapetrou [64]. Notice that there are six functions of the coordinates t and r, while the θ dependence is fixed by the symmetry. Two out of the six functions vanish whenever one restricts to symmetric tensors, that is, $g(X, Y) = g(Y, X)$, such as the metric tensor. The form of the symmetric, covariant, rank-two tensor is

$$g = \begin{pmatrix} A(t,r) & B(t,r) & 0 & 0 \\ B(t,r) & C(t,r) & 0 & 0 \\ 0 & 0 & D(t,r) & 0 \\ 0 & 0 & 0 & D(t,r)\sin^2(\theta) \end{pmatrix}, \tag{18}$$

which, under a redefinition of the radial and temporal coordinates, takes the standard form:

$$g = \begin{pmatrix} F(t,r) & 0 & 0 & 0 \\ 0 & G(t,r) & 0 & 0 \\ 0 & 0 & D(t,r) & 0 \\ 0 & 0 & 0 & D(t,r)\sin^2(\theta) \end{pmatrix}. \tag{19}$$

It is worth noticing that Eq. (19) is the most general spherical ansatz which is the one wants to solve Einstein's field equations. The static condition is only assured by the Birkhoff theorem [65–68], once the field equations are given.

5.1.2. Isotropic affine connection

The strategy used in the previous section can be repeated for an affine connection, and we shall end up with the most general isotropic (affine) connection. For the sake of simplicity, we do not include the differential equations obtained from the calculation of the Lie derivative.[5]

As before, the Lie derivative along the generators of the spherical symmetry fixes the angular dependence of the connection's components. The nonvanishing components of an isotropic $\widehat{\Gamma}^a_{bc}$, are

$$\begin{array}{ll}
\widehat{\Gamma}^t_{tt} = F_{000}(t,r) & \widehat{\Gamma}^t_{tr} = F_{001}(t,r) \\
\widehat{\Gamma}^t_{rt} = F_{010}(t,r) & \widehat{\Gamma}^t_{rr} = F_{011}(t,r) \\
\widehat{\Gamma}^t_{\theta\theta} = F_{033}(t,r) & \widehat{\Gamma}^t_{\theta\phi} = F_{023}(t,r)\sin(\theta) \\
\widehat{\Gamma}^t_{\phi\theta} = -F_{023}(t,r)\sin(\theta) & \widehat{\Gamma}^t_{\phi\phi} = F_{033}(t,r)\sin^2(\theta) \\
\widehat{\Gamma}^r_{tt} = F_{100}(t,r) & \widehat{\Gamma}^r_{tr} = F_{101}(t,r) \\
\widehat{\Gamma}^r_{rt} = F_{110}(t,r) & \widehat{\Gamma}^r_{rr} = F_{111}(t,r) \\
\widehat{\Gamma}^r_{\theta\theta} = F_{133}(t,r) & \widehat{\Gamma}^r_{\theta\phi} = F_{123}(t,r)\sin(\theta) \\
\widehat{\Gamma}^r_{\phi\theta} = -F_{123}(t,r)\sin(\theta) & \widehat{\Gamma}^r_{\phi\phi} = F_{133}(t,r)\sin^2(\theta)
\end{array}$$

---

[5] Notice that in four dimensions, an affine connection has 64 components. Therefore, there are 192 differential equations to solve for the 3 generators of spherical symmetry.

$$
\begin{aligned}
&\widehat{\Gamma}^{\theta}_{t\theta} = F_{303}(t,r) && \widehat{\Gamma}^{\theta}_{t\phi} = -F_{302}(t,r)\sin(\theta) \\
&\widehat{\Gamma}^{\theta}_{r\theta} = F_{313}(t,r) && \widehat{\Gamma}^{\theta}_{r\phi} = -F_{312}(t,r)\sin(\theta) \\
&\widehat{\Gamma}^{\theta}_{\theta t} = F_{330}(t,r) && \widehat{\Gamma}^{\theta}_{\theta r} = F_{331}(t,r) \\
&\widehat{\Gamma}^{\theta}_{\phi t} = -F_{320}(t,r)\sin(\theta) && \widehat{\Gamma}^{\theta}_{\phi r} = -F_{321}(t,r)\sin(\theta) \\
&\widehat{\Gamma}^{\theta}_{\phi\phi} = -\cos(\theta)\sin(\theta) && \widehat{\Gamma}^{\phi}_{t\theta} = \frac{F_{302}(t,r)}{\sin(\theta)} \\
&\widehat{\Gamma}^{\phi}_{t\phi} = F_{303}(t,r) && \widehat{\Gamma}^{\phi}_{r\theta} = \frac{F_{312}(t,r)}{\sin(\theta)} \\
&\widehat{\Gamma}^{\phi}_{r\phi} = F_{313}(t,r) && \widehat{\Gamma}^{\phi}_{\theta t} = \frac{F_{320}(t,r)}{\sin(\theta)} \\
&\widehat{\Gamma}^{\phi}_{\theta r} = \frac{F_{321}(t,r)}{\sin(\theta)} && \widehat{\Gamma}^{\phi}_{\theta\phi} = \frac{\cos(\theta)}{\sin(\theta)} \\
&\widehat{\Gamma}^{\phi}_{\phi t} = F_{330}(t,r) && \widehat{\Gamma}^{\phi}_{\phi r} = F_{331}(t,r) \\
&\widehat{\Gamma}^{\phi}_{\phi\theta} = \frac{\cos(\theta)}{\sin(\theta)}.
\end{aligned}
$$

This general isotropic connection depends on 20 functions of the coordinates t and r, and has nonvanishing torsion and nonmetricity. Therefore, for our purposes within this paper, we can restrict even further to a torsion-free connection, whose components are

$$
\begin{aligned}
&\Gamma^{t}_{tt} = F_{000}(t,r) && \Gamma^{t}_{tr} = F_{001}(t,r) \\
&\Gamma^{t}_{rr} = F_{011}(t,r) && \Gamma^{t}_{\theta\theta} = F_{033}(t,r) \\
&\Gamma^{t}_{\phi\phi} = F_{033}(t,r)\sin^2(\theta) && \Gamma^{r}_{tt} = F_{100}(t,r) \\
&\Gamma^{r}_{tr} = F_{101}(t,r) && \Gamma^{r}_{rr} = F_{111}(t,r) \\
&\Gamma^{r}_{\theta\theta} = F_{133}(t,r) && \Gamma^{r}_{\phi\phi} = F_{133}(t,r)\sin^2(\theta) \\
&\Gamma^{\theta}_{t\theta} = F_{303}(t,r) && \Gamma^{\theta}_{t\phi} = -F_{302}(t,r)\sin(\theta) \\
&\Gamma^{\theta}_{r\theta} = F_{313}(t,r) && \Gamma^{\theta}_{r\phi} = -F_{312}(t,r)\sin(\theta) \\
&\Gamma^{\theta}_{\phi\phi} = -\cos(\theta)\sin(\theta) && \Gamma^{\phi}_{t\theta} = \frac{F_{302}(t,r)}{\sin(\theta)} \\
&\Gamma^{\phi}_{t\phi} = F_{303}(t,r) && \Gamma^{\phi}_{r\theta} = \frac{F_{312}(t,r)}{\sin(\theta)} \\
&\Gamma^{\phi}_{r\phi} = F_{313}(t,r) && \Gamma^{\phi}_{\theta\phi} = \frac{\cos(\theta)}{\sin(\theta)}.
\end{aligned}
\tag{20}
$$

The last connection depends on 12 functions of t and r, and these are the functions to be fixed by solving the field Eq. (4).

Using Eq. (20), we calculated the Ricci tensor and noticed that it is not symmetric, because the connection is not equi-affine. In order for the Ricci to be symmetric, the following conditions must hold:

$$\frac{\partial F_{000}}{\partial r} - \frac{\partial F_{001}}{\partial t} + 2\frac{\partial F_{303}}{\partial r} = 0, \tag{21}$$

and

$$\frac{\partial F_{101}}{\partial r} - \frac{\partial F_{101}}{\partial t} + 2\frac{\partial F_{313}}{\partial r} = 0. \tag{22}$$

There are several solutions to these conditions, and each of them could (in principle) provide a solution to the field Eq. (4). It is worth mentioning that if we do not demand the connection to be equi-affine, the field equations to solve would be Eq. (3), which has an extra term which cannot be obtained from the Yang-Mills-like effective action in Eq. (8).

5.2. Cosmological ansätze

The Lorentian isotropic and homogeneous spaces in four dimensions have isometry group either $SO(4)$, $SO(3,1)$ or $ISO(3)$, and their algebra can be obtained from the algebra $\mathfrak{so}(4)$ through a 3+1 decomposition, that is, $J_{AB}=\{J_{ab}, J_{a*}\}$,[6] where the extra dimension has been denoted by an asterisk. In terms of these new generators, the algebra reads

$$\begin{aligned}
[J_{ab}, J_{cd}] &= \delta_{bc}J_{ad} - \delta_{ac}J_{bd} + \delta_{ad}J_{bc} - \delta_{bd}J_{ac}, \\
[J_{ab}, J_{c*}] &= \delta_{bc}J_{a*} - \delta_{ac}J_{b*}, \\
[J_{a*}, J_{c*}] &= -\kappa J_{ac},
\end{aligned} \tag{23}$$

with[7]

$$\kappa = \begin{cases} 1 & SO(4) \\ 0 & ISO(3) \\ -1 & SO(3,1) \end{cases}. \tag{24}$$

One can express the Killing vectors in spherical coordinates, and in addition to those obtained in Eq. (15), we get

[6] We shall use the standard (Euclidean) three-dimensional correspondence $J_a = \frac{1}{2}\varepsilon_{abc}J^{bc}$, in order to match notations with the previous case.

[7] The inhomogeneous algebra of $ISO(n)$ can be obtained from the ones of $SO(n+1)$ or $SO(n,1)$ through the Inönü Wigner contraction [69].

http://dx.doi.org/10.5772/intechopen.70951

$$
\begin{aligned}
P_1 = J_{1*} &= \sqrt{1-\kappa r^2}\begin{pmatrix} 0 & \cos(\varphi)\sin(\theta) & \dfrac{\cos(\varphi)\cos(\theta)}{r} & -\dfrac{\sin(\varphi)}{r\sin(\theta)} \end{pmatrix}, \\
P_2 = J_{2*} &= \sqrt{1-\kappa r^2}\begin{pmatrix} 0 & \sin(\varphi)\sin(\theta) & \dfrac{\cos(\theta)\sin(\varphi)}{r} & \dfrac{\cos(\varphi)}{r\sin(\theta)} \end{pmatrix}, \\
P_3 = J_{3*} &= \sqrt{1-\kappa r^2}\begin{pmatrix} 0 & \cos(\theta) & -\dfrac{\sin(\theta)}{r} & 0 \end{pmatrix}.
\end{aligned} \tag{25}
$$

5.2.1. *Isotropic and homogeneous (covariant)-two tensor*

In order to find the most general isotropic and homogeneous covariant two tensor, we can start from the result in Eq. (17) and impose now the symmetries from the extra generators. The equations are[8]

$$
\begin{aligned}
&£_{P_3}T_{tt} : \sqrt{1-\kappa r^2}\cos(\theta)\frac{\partial T_{00}}{\partial r} = 0, && £_{P_3}T_{tr} : \left(\kappa r T_{01} + (\kappa r^2 - 1)\frac{\partial T_{01}}{\partial r}\right) = 0, \\
&£_{P_3}T_{t\theta} : T_{01} = 0, && £_{P_3}T_{rt} : \left(\kappa r T_{10} + (\kappa r^2 - 1)\frac{\partial T_{10}}{\partial r}\right) = 0, \\
&£_{P_3}T_{rr} : \left(2\kappa r T_{11} + (\kappa r^2 - 1)\frac{\partial T_{11}}{\partial r}\right) = 0, && £_{P_3}T_{r\theta} : (\kappa r^4 - r^2)T_{11} + T_{22} = 0, \\
&£_{P_3}T_{r\varphi} : H = 0, && £_{P_3}T_{\theta t} : T_{10} = 0, \\
&£_{P_3}T_{\theta r} : (\kappa r^4 - r^2)T_{11} + T_{22} = 0, && £_{P_3}T_{\theta\theta} : \left(r\frac{\partial T_{22}}{\partial r} - 2T_{22}\right) = 0, \\
&£_{P_3}T_{\theta\varphi} : \left(r\frac{\partial H}{\partial r} - 2H\right) = 0, && £_{P_3}T_{\varphi r} : H = 0, \\
&£_{P_3}T_{\varphi\theta} : \left(r\frac{\partial H}{\partial r} - 2H\right) = 0, && £_{P_3}T_{\varphi\varphi} : \left(r\frac{\partial T_{22}}{\partial r} - 2T_{22}\right) = 0.
\end{aligned} \tag{26}
$$

The above equations are solved for a tensor of the form:

$$
T = G_{00}(t)\mathrm{d}t\otimes\mathrm{d}t + \frac{G_{11}(t)}{1-\kappa r^2}\mathrm{d}r\otimes\mathrm{d}r + r^2G_{11}(t)\mathrm{d}\theta\otimes\mathrm{d}\theta + r^2G_{11}(t)\sin^2(\theta)\mathrm{d}\varphi\otimes\mathrm{d}\varphi. \tag{27}
$$

Under a redefinition of the "time" coordinate, the tensor is nothing but the standard ansatz for the Friedman-Robertson-Walker metric.

5.2.2. *Isotropic and homogeneous affine connection*

Without further details, we present the nonvanishing components of an isotropic and homogeneous affine connection. It is given by

[8] Due to the isotropy, one needs to solve just for one of these extra generators.

$$
\begin{aligned}
\widehat{\Gamma}^{t}_{tt} &= G_{000}(t) & \widehat{\Gamma}^{t}_{rr} &= \frac{G_{011}(t)}{1-\kappa r^2} \\
\widehat{\Gamma}^{t}_{\theta\theta} &= r^2 G_{011}(t) & \widehat{\Gamma}^{t}_{\varphi\varphi} &= r^2 G_{011}(t)\sin^2(\theta) \\
\widehat{\Gamma}^{r}_{tr} &= G_{101}(t) & \widehat{\Gamma}^{r}_{rt} &= G_{110}(t) \\
\widehat{\Gamma}^{r}_{rr} &= \frac{\kappa r}{1-\kappa r^2} & \widehat{\Gamma}^{r}_{\theta\theta} &= \kappa r^3 - r \\
\widehat{\Gamma}^{r}_{\theta\varphi} &= \sqrt{1-\kappa r^2}\, r^2 G_{123}(t)\sin(\theta) & \widehat{\Gamma}^{r}_{\varphi\theta} &= -\sqrt{1-\kappa r^2}\, r^2 G_{123}(t)\sin(\theta) \\
\widehat{\Gamma}^{r}_{\varphi\varphi} &= \left(\kappa r^3 - r\right)\sin^2(\theta) & \widehat{\Gamma}^{\theta}_{t\theta} &= G_{101}(t) \\
\widehat{\Gamma}^{\theta}_{r\theta} &= \frac{1}{r} & \widehat{\Gamma}^{\theta}_{r\varphi} &= -\frac{G_{123}(t)\sin(\theta)}{\sqrt{1-\kappa r^2}} \\
\widehat{\Gamma}^{\theta}_{\theta t} &= G_{110}(t) & \widehat{\Gamma}^{\theta}_{\theta r} &= \frac{1}{r} \\
\widehat{\Gamma}^{\theta}_{\varphi r} &= \frac{G_{123}(t)\sin(\theta)}{\sqrt{1-\kappa r^2}} & \widehat{\Gamma}^{\theta}_{\varphi\varphi} &= -\cos(\theta)\sin(\theta) \\
\widehat{\Gamma}^{\varphi}_{t\varphi} &= G_{101}(t) & \widehat{\Gamma}^{\varphi}_{r\theta} &= \frac{G_{123}(t)}{\sqrt{1-\kappa r^2}\sin(\theta)} \\
\widehat{\Gamma}^{\varphi}_{r\varphi} &= \frac{1}{r} & \widehat{\Gamma}^{\varphi}_{\theta r} &= -\frac{G_{123}(t)}{\sqrt{1-\kappa r^2}\sin(\theta)} \\
\widehat{\Gamma}^{\varphi}_{\theta\varphi} &= \frac{\cos(\theta)}{\sin(\theta)} & \widehat{\Gamma}^{\varphi}_{\varphi t} &= G_{110}(t) \\
\widehat{\Gamma}^{\varphi}_{\varphi r} &= \frac{1}{r} & \widehat{\Gamma}^{\varphi}_{\varphi\theta} &= \frac{\cos(\theta)}{\sin(\theta)},
\end{aligned}
\tag{28}
$$

which is determined by five independent functions, but this affine connection still possess torsion. The imposition of vanishing torsion kills two of the above functions, and the remaining components of the connection are

$$
\begin{aligned}
\Gamma^{t}_{tt} &= G_{000}(t) & \Gamma^{t}_{rr} &= \frac{G_{011}(t)}{1-\kappa r^2} \\
\Gamma^{t}_{\theta\theta} &= r^2 G_{011}(t) & \Gamma^{t}_{\varphi\varphi} &= r^2 G_{011}(t)\sin^2(\theta) \\
\Gamma^{r}_{tr} &= G_{101}(t) & \Gamma^{r}_{rr} &= \frac{\kappa r}{1-\kappa r^2} \\
\Gamma^{r}_{\theta\theta} &= \kappa r^3 - r & \Gamma^{r}_{\varphi\varphi} &= \left(\kappa r^3 - r\right)\sin^2(\theta) \\
\Gamma^{\theta}_{t\theta} &= G_{101}(t) & \Gamma^{\theta}_{r\theta} &= \frac{1}{r} \\
\Gamma^{\theta}_{\varphi\varphi} &= -\cos(\theta)\sin(\theta) & \Gamma^{\varphi}_{t\varphi} &= G_{101}(t) \\
\Gamma^{\varphi}_{r\varphi} &= \frac{1}{r} & \Gamma^{\varphi}_{\theta\varphi} &= \frac{\cos(\theta)}{\sin(\theta)}.
\end{aligned}
\tag{29}
$$

6. Towards the solution of the field equations

6.1. Cosmological solutions

Using the connection in Eq. (29), the Ricci is calculated and yields

$$\begin{aligned} R_{tt} &= 3G_{000}(t)G_{101}(t) - 3G_{101}(t)^2 - 3\frac{\partial G_{101}}{\partial t} \\ R_{rr} &= \frac{G_{000}(t)G_{011}(t) + G_{011}(t)G_{101}(t) + 2\kappa + \frac{\partial G_{011}}{\partial t}}{1-\kappa r^2} \\ R_{\theta\theta} &= \left(G_{000}(t)G_{011}(t) + G_{011}(t)G_{101}(t) + 2\kappa + \frac{\partial G_{011}}{\partial t}\right) r^2 \\ R_{\varphi\varphi} &= \left(G_{000}(t)G_{011}(t) + G_{011}(t)G_{101}(t) + 2\kappa + \frac{\partial G_{011}}{\partial t}\right) r^2 \sin^2(\theta). \end{aligned} \tag{30}$$

Notice that the Ricci tensor is symmetric, determined by only three independent functions, and as expected by the symmetry, just two of their components behave differently.

A first kind of solutions can be found by solving the system of equations determined by vanishing Ricci. However, this strategy requires the fixing of one of the unknown functions. A solution inspired in the components of the connection for Friedmann-Robertson-Walker gives

$$G_{000} = 0, G_{101} = \frac{1}{t - C_1}, G_{011} = -2\kappa t + C_2. \tag{31}$$

A second class of solutions can be found by solving the parallel Ricci equation, $\nabla_\lambda R_{\mu\nu}=0$, which surprisingly yields three independent field equations:

$$\nabla_t R_{tt} = -6\, G_{000}^2 G_{101} + 6\, G_{000} G_{101}^2 + 3\, G_{101}\frac{\partial G_{000}}{\partial t} + 3\,(3\, G_{000} - 2\, G_{101})\frac{\partial G_{101}}{\partial t} - 3\,\frac{\partial^2 G_{101}}{\partial t^2}, \tag{32}$$

$$\nabla_t R_{ii} \sim 2\, G_{011} G_{101}^2 + 2\,(G_{000}G_{011} + 2\,\kappa)G_{101} - G_{011}\frac{\partial G_{000}}{\partial t}, \tag{33}$$

$$-(G_{000} - G_{101})\frac{\partial G_{011}}{\partial t} - G_{011}\frac{\partial G_{101}}{\partial t} - \frac{\partial^2 G_{011}}{\partial t^2}$$

$$\nabla_i R_{ti} = 2\, G_{011} G_{101}^2 - 2\,(2\, G_{000} G_{011} + \kappa)G_{101} - G_{101}\frac{\partial G_{011}}{\partial t} + 3\, G_{011}\frac{\partial G_{101}}{\partial t}. \tag{34}$$

However, the system of equations is complicated enough to avoid an analytic solution.[9] Despite the complication, we can try a couple of assumptions that simplify the system of equations, for example:

[9] Of course one can propose a formal solution in terms of power series. However, this approach that has been used in a paper is still under developed.

- Again, inspired in the Friedmann-Robertson-Walker results, we choose $G_{000}=0$, and solve for the other functions (see [70]).
- We could choose $G_{000}=G_{101}$, to eliminate the nonlinear terms in Eq. (32).
- We can solve both Eqs. (10) and (34) by setting $G_{011}=0$, $\kappa\neq 0$ and (say) $G_{101}=\text{const.}$, and solve Eq. (32) which turns to be a known ordinary differential equation.

Finally, the third class of solutions is that of Eq. (4). The set of equations degenerate and yield a single independent field equation:

$$4\,G_{011}G_{101}^2 - 2\,(G_{000}G_{011} - \kappa)G_{101} - G_{011}\frac{\partial G_{000}}{\partial t} - G_{000}\frac{\partial G_{011}}{\partial t} + 2\,G_{011}\frac{\partial G_{101}}{\partial t} - \frac{\partial^2\,G_{011}}{\partial t^2} = 0. \quad (35)$$

Therefore, we need to set two out of the three unknown functions to be able to solve the connection.

7. Conclusions

We have presented a short review of the polynomial affine gravity, whose field equations (in the torsion-free sector) generalise those of the standard general relativity. In the mentioned approximation, the field equations coincide with (part of) those of the gravitational Yang-Mills theory of gravity, known in the literature as the SKY model.

Among the features of the polynomial affine gravity, we highlighted the following:

- Although our spacetime could be metric, the metric plays no role in the model building.
- The nonrelativistic limit of the model yields a Keplerian potential, even with the contributions of torsion and nonmetricity.
- In the torsion-free sector, all the vacuum solutions to the Einstein gravity are solutions of the polynomial affine gravity.
- Scalar matter can be coupled to the polynomial affine gravity through a symmetric, $\binom{2}{0}$-tensor density. The coupled field equations can be written in a similar form to those of general relativity, with the subtlety that wherever the metric appears in Einstein's equations, we just need a covariantly constant, symmetric, $\binom{0}{2}$-tensor.
- Although the model is built up without the necessity of a metric, one can still assume that the connection is a metric potential. Such consideration yields to obtain new solutions to Eq. (4), which are not solutions of the standard Einstein's equations.[10]

[10] Some solutions will be presented in [70].

- We found the general ansatz for the connection, compatible with isotropic and cosmological symmetries. Additionally, we have sketched the road to solving Eq. (4), for the cosmological connection ansatz. A thorough analysis will be presented in [70].

We would like to finish commenting about the necessity of coupling other forms of matters and a formal counting of degrees of freedoms.

Acknowledgements

O.C-F. wants to thank to the ICTP-SAIFR and IFT-UNESP (Sao Paulo, Brazil) for the hospitality while finishing this article. This work has been partially founded by the CONICYT (Chile) project PAI-79140040. The Centro Científico Tecnológico de Valparaíso (CCTVal) is funded by the Chilean government through the Centres of Excellence Basal Financing Program FB0821 of CONICYT.

Author details

Oscar Castillo-Felisola[1,2*]

*Address all correspondence to: o.castillo.felisola@gmail.com

1 Centro Científico Tecnológico de Valparaíso, Valparaíso, Chile

2 Departamento de Física, Universidad Técnica Federico Santa María, Valparaíso, Chile

References

[1] Einstein A. Zur allgemeinen relativitätstheorie. Sitzungsberichte der Preussischen Akademie der Wissenschaften. 1915;**1**:778

[2] Hilbert D. Die grundlagen der physik.(erste mitteilung). Nachrichten von der Gesellschaft der Wissenschaften zu Göttingen, Mathematisch-Physikalische Klasse. 1915;**1915**:395

[3] Hooft G 't. An algorithm for the poles at dimension four in the dimensional regularization procedure. Nuclear Physics B. 1973;**62**:444

[4] Hooft G 't, Veltman MJG. One loop divergencies in the theory of gravitation. Annales de l'Institut Henri Poincare, Section A: Physique Theorique. 1974;**20**:69

[5] Deser S, van Nieuwenhuizen P. One loop divergences of quantized Einstein-Maxwell fields. Physical Review D. 1974;**10**:401

[6] Deser S, van Nieuwenhuizen P. Nonrenormalizability of the quantized Dirac-Einstein system. Physical Review D. 1974;**10**:411

[7] DeWitt BS. Quantum theory of gravity. 1. The canonical theory. Physics Review. 1967; **160**:1113

[8] DeWitt BS. Quantum theory of gravity. 2. The manifestly covariant theory. Physics Review. 1967;**162**:1195

[9] DeWitt BS. Quantum theory of gravity. 3. Applications of the covariant theory. Physics Review. 1967;**162**:1239

[10] Rovelli C. Notes for a brief history of quantum gravity. https://arxiv.org/abs/gr-qc/0006061. 2000

[11] Eddington AS. The Mathematical Theory of Relativity. Cambridge: Cambridge University Press. 1923

[12] Schrödinger E. Space-Time Structure. Cambridge University Press; 1950

[13] Cartan E. Sur une généralisation de la notion de courbure de riemann et les espaces à torsion. Comptes Rendus de l'Acad emie des Sciences. Paris. 1922;**174**:593

[14] Cartan E. Sur les variétés à connexion affine et la théorie de la relativité généralisée (première partie). Annales Scientifiques de l'Ecole Normale Superieure. 1923;**40**:325

[15] Cartan E. Sur les variétés à connexion affine, et la théorie de la relativité généralisée (première partie) (suite). Annales Scientifiques de l'Ecole Normale Superieure. 1924;**41**:1

[16] Cartan E. Sur les variétés à connexion affine et la théorie de la relativité généralisée, (deuxième partie). Annales Scientifiques de l'Ecole Normale Superieure. 1925;**42**:17

[17] Kibble TWB. Lorentz invariance and the gravitational field. Journal of Mathematical Physics. 1961;**2**:212-221

[18] Sciama DW. The physical structure of general relativity. Reviews of Modern Physics. 1964;**36**:463

[19] Krasnov K. Non-metric gravity: A status report. Modern Physics Letters A. 2007;**22**:3013-3026

[20] Plebanski JF. On the separation of Einsteinian substructures. Journal of Mathematical Physics. 1977;**18**:2511

[21] Kijowski J. On a new variational principle in general relativity and the energy of the gravitational field. General Relativity and Gravitation. 1978;**9**(10):857

[22] Ferraris M, Kijowski J. General relativity is a gauge type theory. Letters in Mathematical Physics. 1981;**5**(2):127

[23] Ferraris M, Kijowski J. On the equivalence of the relativistic theories of gravitation. General Relativity and Gravitation. 1982;**14**(2):165

[24] Krasnov K. Non-metric gravity. I. Field equations. Classical and Quantum Gravity. 2008; **25**:025001

[25] Krasnov K, Shtanov Y. Non-metric gravity. II. Spherically symmetric solution, missing mass and redshifts of quasars. Classical and Quantum Gravity. 2008;**25**:025002

[26] Krasnov K. Pure connection action principle for general relativity. Physical Review Letters. 2011;**106**:251103

[27] Nikodem J. Popławski. On the nonsymmetric purely affine gravity. Modern Physics Letters A. 2007;**22**(36):2701

[28] Nikodem J. Popławski. Affine theory of gravitation. General Relativity and Gravitation, **46**:1625, 2014

[29] Castillo-Felisola O, Skirzewski A. A polynomial model of purely affine Gravity. Revista Mexicana de Fisica. 2015;**61**:421

[30] Castillo-Felisola O and Skirzewski A. Einstein's gravity from an affine model. https://arxiv.org/abs/1505.04634. 2015

[31] Hehl FW, Dermott McCrea J, Mielke EW, Yuval N'e. Metric affine gauge theory of gravity: Field equations, noether identities, world spinors, and breaking of dilation invariance. Physics Reports. 1995;**258**:1-171

[32] Schouten JA. Ricci-calculus: An introduction to tensor analysis and its geometrical applications. Vol. 10. Berlin: Springer. 2013

[33] Ortín T. Gravity and Strings, Cambridge Monographs on Mathematical Physics. 2nd ed. Vol. 5. Cambridge: Cambridge University Press. 2015

[34] Nikodem J. Popławski. Cosmological constant from quarks and torsion. Annals of Physics, **523**:291, 2011

[35] Nikodem J. Popławski. Four-fermion interaction from torsion as dark energy. General Relativity and Gravitation. 2012;**44**:491–499

[36] Nikodem J. Popławski. Cosmology with torsion—An alternative to cosmic inflation. Physics Letters B. 2010;**694**:181

[37] Castillo-Felisola O, Corral C, del Pino S, Ramírez F. Kaluza–Klein cosmology from five-dimensional Lovelock–Cartan theory. Physical Review D. 2016;**94**(12):124020

[38] Belyaev AS, Shapiro IL. Torsion action and its possible observables. Nuclear Physics B. 1999;**543**:20

[39] Belyaev AS, Shapiro IL, do Vale MAB. Torsion phenomenology at the LHC. Physical Review D. 2007;**75**:034014

[40] Fabbri L, Vignolo S. A modified theory of gravity with torsion and its applications to cosmology and particle physics. International Journal of Theoretical Physics. 2012;**51**: 3186-3207

[41] Fabbri L. A Torsional Model of Leptons. Modern Physics Letters A. 2012;**A27**:1250199

[42] Capozziello S, Fabbri L, Vignolo S. Weak Forces and neutrino oscillations under the standards of hybrid gravity with torsion. Modern Physics Letters A. 2013;**28**(35):1350155

[43] Castillo-Felisola O, Corral C, Villavicencio C, Zerwekh AR. Fermion masses through condensation in spacetimes with torsion. Physical Review D. 2013;**88**:124022

[44] Castillo-Felisola O, Corral C, Schmidt I, Zerwekh AR. Updated limits on extra dimensions through torsion and LHC data. Modern Physics Letters A. 2014;**29**:1450081

[45] Castillo-Felisola O, Corral C, Kovalenko S, Schmidt I. Torsion in extra dimensions and one-loop observables. Physical Review D. 2014;**90**:024005

[46] Castillo-Felisola O, Corral C, Kovalenko S, Schmidt I, Lyubovitskij VE. Axions in gravity with torsion. Physical Review D. 2015;**91**(8):085017

[47] Curtright T. Generalized gauge fields. Physics Letters B. 1985;**165**:304

[48] McGady DA, Rodina L. Higher-spin massless *S*-matrices in four-dimensions. Physical Review D. 2014;**90**(8):084048

[49] Camanho XO, Edelstein JD, Maldacena J, Zhiboedov A. Causality constraints on corrections to the graviton three-point coupling. Journal of High Energy Physics. 2016 Feb 1; **2016**(2):20

[50] Nomizu K, Sasaki T. Affine differential geometry. Cambridge: Cambridge University Press. 1994

[51] Robert L. Bryant. Symmetries of non-Riemannian curvature tensor (answer). MathOverflow. Available from: http://mathoverflow.net/a/212794/25356 (Accessed: 3 August 2015)

[52] Bourguignon J-P. Les variétés de dimension 4 à signature non nulle dont la courbure est harmonique sont d'einstein. Inventiones Mathematicae. 1981;**63**(2):263

[53] Bourguignon J-P, Blaine Lawson H Jr. Yang-Mills theory: Its physical origins and differential geometric aspects. In: Seminar on differential Geometry, Annals of Mathematics Studies. Vol. 102. 1982. p. 395

[54] Nakahara M. Geometry, Topology and Physics. London: Institute of Physics. 2005

[55] Stephenson G. Quadratic lagrangians and general relativity. Nuovo Cimento. 1958;**9**(2): 263-269

[56] Kilmister CW, Newman DJ. The use of algebraic structures in physics. In: Mathematical Proceedings of the Cambridge Philosophical Society. Vol. 57. Cambridge: Cambridge University Press. 1961. p. 851

[57] Yang CN. Integral formalism for gauge fields. Physical Review Letters. Aug 1974;**33**:445

[58] Bekenstein JD, Majhi BR. Is the principle of least action a must? Nuclear Physics B. 2015;**892**:337

[59] Yano K. The theory of Lie derivatives and its applications, volume 3. Amsterdam: North-Holland. 1957

[60] Choquet-Bruhat Y, DeWitt-Morette C, Dillard-Bleick M. Analysis, manifolds and physics. Vol. 1 & 2. Amsterdam: North-Holland. 1989

[61] McInerney A. First Steps in Differential Geometry. New York: Springer. 2013

[62] Stein WA, et al. Sage Mathematics Software (Version 7.6). The Sage Development Team; 2017

[63] Gourgoulhon E, Bejger M, et al. SageManifolds (Version 1.0.1). SageManifolds Development Team; 2017

[64] Papapetrou A. Static spherically symmetric solutions in the unitary field theory. Proceedings of the Royal Irish Academy A. 1948;**52**:69

[65] Jebsen JT. On the General Spherically Symmetric Solutions of Einstein's Gravitational Equations in Vacuo. Arkiv for Matematik, Astronomi och Fysik. 1921;**15**:18

[66] Birkhoff GD. Relativity and Modern Physics. Boston: Harvard University Press. 1923

[67] Alexandrow W. Über den kugelsymmetrischen Vakuumvorgang in der Einsteinschen Gravitationstheorie. Annalen der Physik. 1923;**377**(18):141

[68] Eiesland J. The group of motions of an Einstein space. Transactions of the American Mathematical Society. 1925;**27**(2):213

[69] Gilmore R. Lie Groups, Lie Algebras, and Some of their Applications. New York: Dover. 2005

[70] Oscar Castillo-Felisola, Oscar Orellana, and Aureliano Skirzewski. Metric and non-metric solutions to the polynomial affine gravity (in preparation)

Chapter 10

Gravity, Curvature and Energy: Gravitational Field Intentionality to the Cohesion and Union of the Universe

Francisco Bulnes

Additional information is available at the end of the chapter

http://dx.doi.org/10.5772/intechopen.71037

Abstract

We use the quantum operators O_c^G, which are diffeomorphisms of gravity creating the intentionality under the action integrals to prove and determine the gluing intention for adherence of the matter-energy (taking the corresponding mass-energy tensor T^{ab}) to create complex bodies in the scale of conforming the fragmented Universe such as we know. The reverse is the planting of the energy model of gravity in accordance with the implications in space-time due to the diffeomorphisms of gravity, which were designed to explain the existence of the intention as kernel of the integral operators of the actions with this intention as direction of the energy-matter. The time, in particular, can be shown through instantons of a gauge field (this as electromagnetic field, and in this case appears the torsion) of gravity, which appears in natural way as the same integral operators obtained. Finally, using the complex Riemannian structure of our model of the space-time, and the K-invariant G-structure of the orbits used to obtain curvature, are obtained as consequences of the diffeomorphisms, the field equations to the energy-matter tensor density in each case of the gravitational field.

Keywords: agglutination and gluing of sticking cells, curvature on quantum gravity, gravitational field intention, gravitational diffeomorphisms, integration invariants, local diffeomorphisms, regular representation

1. Introduction

The Universe outside their quantum model is a fragmented cosmos, where the matter is separated in different gradients of interaction of particles of fundamental level, which goes conforming scales of the matter until to shape the complex structures of all sidereal objects and the life that we know. However, in this process and for the alone presence of matter in the space is created a permanent will with persistent action and the long scope which is the gravitational field.

The permanence of this field is the control aspects of regulation in the sidereal objects such as the movements, shape of the objects, shape of the local space, auto-gravitation, and conformation of the chemical structures. Also their self-recomposing, as well as the different natural process to the life as the kinematic equilibrium of the living beings, biorhythm, biochemical auto-regulations, etcetera, which requires sedimentation and other process that brings the gravitational field, as example, the connection in equilibrium levels depending of the position between all bodies in the space. In addition, the time is a consequence possible of the gravitational field which has an effect, the curvature, where the time being the distance between cause and effect; the instantons [1] [1] of the curvature energy is the distance between gravitational field and curvature in layers of the matter-energy that goes composing the sidereal objects and the Universe. Likewise, this determines an integrated action with the intention of connect the space-time with gravity as the encoding of the Universe, at least in the Einstein-Cartan field conception considering also the Dirac equation to the energy level.

Likewise, we have:

Conjecture 1.1 (F. Bulnes). Gravity tends to join and agglutinate. Their intention is cohesion and shape of the Universe.

Why the gravity intention is the cohesion? Because, in the beginning of the Universe, the only ingredient was energy. After the Big Bang, the energy was being fragmented in matter and other energy manifestations in the Universe, such as we know in our days. The gravity is the ingredient of that primal energy (the energy before the Big Bang), but this is the version of that primal energy when matter in the space and time appeared (as well as local time or global time in the Universe).

Why the gravity intention is the shape? Because the Universe needs the gravity, to be used in the interpretations of elements of the Universe as are time, space, and energy. For example, space could be curvature, time could be second curvature (or torsion), and energy could be curvature energy in any of their modalities (first and second curvature energies).

2. Integrated action of gravity with the space-time

The Universe before their fragmentation (as has been interpreted in different cosmological theories in the beginning of the Universe) was conformed for energy due to the Big Bang. After when this was freeze gradually is conformed for a quantum process in an Early Universe composed in their totality by neutrinos as basic particles components. These neutrinos were differentiated for a natural process in new neutrinos giving some aspects of our Universe, as possibly the dark matter production. Other neutrino classes as the right-handed neutrinos were establishing with the baryon-genesis process; and other particles as the tachyons and

[1] An instanton, or also called pseudo-particle in particle physics, is in theoretical and mathematical physics, a classical solution to motion equations with a finite, non-null action as well as quantum mechanics or in quantum field theory. Likewise, this solution is a solution to the equations of motion in classical field theory on a Euclidean space-time.

the fermions/anti-fermions in a polarization process creating the charge $Q[k]$, for one side, and for other side, with the mass particles that we know. Likewise, using theories that involve the spinor theories, massive objects in the Universe (modeled this as complex Riemannian manifold) can be detected and measured. This is given for the perturbation created in the space-time near of the presence of this object [2], when the gauge particle (in this case a photon) passes near the post-Newtonian gravitational limit of the event horizon of the object. The measure needs to be realized in the asymptotic region tending to the flat space far of the massive object; since the photon feels the gravity of the object in the condition of energy since as particle can do it.

To establish the concept of agglutination and join of matter, first it is necessary to analyze the matter on their equivalence, the energy. Before, the Universe was only energy that was available, including the space and time. There was no space and time as we know. These were born after the Big Bang, when the quantum proliferation of particles begun, which conformed the matter particles, these atoms and after molecules. But, what determines this way to create matter. This requires an implicit conscience to create matter from microscopic level, and as this primal matter was part of an energy flow that defines space-time, the rest was stayed as matter energy but without the qualities of compactification of the mass. This gives the beginning to the gravitational energy, which is as natural as the manifestation of a field, the gravitational field.

But, what happens with the space and time? How the gravity enters in the shape of the space and the time? Is gravity consequence of the intention of field in the Early Universe to control the inflation and of this way, to get a defined expansion with a causality-defined arrow?

The torsion of the space is related with the electromagnetic phenomena of the space-time, and gravity more torsion can to define helicities, spins, and polarization rules between particles in the fundamental context of the matter, movement, and space.

Such seems that the intentionality of the field in all the ages of the Universe has been maintaining the cohesion of the space-time such that the primordial component of field, which could be their connection will not be broken and the evolution obeys a sequence of microscopic model (Dirac model) of the Universe, where particles as neutrinos are re-combined. The Lagrangian model must involve the individual actions of gravity and torsion and their joint action,

$$L = \mathrm{L}_G + L_T + L_{\mathrm{GT}}, \tag{1}$$

then the action involving the integral term of gravity and torsion is [3]:

$$\begin{aligned}\mathfrak{I}_{\mathrm{TOTAL}} &= \frac{\mathrm{i}}{2\kappa}\int \mathrm{d}^4 x e e_{\mathrm{a}}^{\mu} e_{\mathrm{b}}^{\nu} \mathrm{R}_{\mu\nu}^{\mathrm{ab}}(\varpi) + \\ &\frac{\mathrm{i}}{2}\int \mathrm{d}^4 x e\left(e_{\mathrm{a}}^{\mu}\overline{\psi}\tilde{\gamma}^{\mathrm{a}}\mathcal{D}_{\mu}\varpi, \mathrm{A})\psi - \overline{\mathcal{D}_{\mu}(\varpi, \mathrm{A})\psi}\tilde{\gamma}^{\mathrm{a}}\psi)\right) + \frac{\mathrm{i}}{2}\int \mathrm{d}^4 x e \frac{3}{16}\,\kappa \mathrm{J}_{(\mathrm{A})}^{\mu}\mathrm{J}_{(\mathrm{A}\mu)},\end{aligned} \tag{2}$$

where $\kappa = 8\pi G_N$, is the gravitational constant that appears in gravitation and $\kappa J_{(A)}^{\mu} J_{(A)\mu}$, is the fermion self-interaction induced by quantum torsion.

The neutrinos and anti-neutrinos, as was mentioned before, are relevant and primordial in the production of matter to obtain gravity. The role of torsion is to explain the mechanisms of matter/anti-matter in gravity.

To establish a Dirac model that gives explanation from the neutrinos and fermions (which establish the energy required to the coupling of torsion and gravity), the joint action between gravity and torsion comes through given torsion connections in the form:

$$\omega_{\mu}^{\mathrm{ab}} = \varpi_{\mu}^{\mathrm{ab}} + \frac{\kappa}{4} \in_{\mathrm{cd}}^{\mathrm{ab}} e_{\mu}^{c} J_{(A)}^{\mu}, \tag{3}$$

However, there exist some ambiguities in the Einstein-Cartan theory with the Immirzi parameter. For example, actions can be modified by total derivatives, without effects to on-shell physics given to create movement (that is to say, equations of motion), but such terms confirm the structural coefficients to the quantum context. For other side, in pure gravity without torsion, one can add the following identically zero term due to Bianchi identities of symmetry properties of the Riemann tensor:

$$\in^{\mu\nu\rho\sigma} \mathrm{R}_{\mu\nu\sigma\rho}(\varpi) = 0, \tag{4}$$

which confirm that the gravity establishes a global behavior in the Universe, not being true in the presence of torsion to the quantum level. For example, the action

$$\mathrm{S}_{\mathrm{Holst}} = -\frac{\beta}{4\kappa} \int \mathrm{d}^4 x e \mathrm{e}_{\mathrm{a}}^{\mu} \mathrm{e}_{\mathrm{b}}^{\mu} \in_{\mathrm{cd}}^{\mathrm{ab}} \mathrm{R}_{\mu\nu}^{\mathrm{cd}}, \tag{5}$$

where the tensor $\tilde{\mathrm{R}}_{\mu\nu}^{\mathrm{ab}} = \in_{\mathrm{cd}}^{\mathrm{ab}} \mathrm{R}_{\mu\nu}^{\mathrm{cd}}$, is the dual of the curvature tensor $\mathrm{R}_{\mu\nu}^{\mathrm{cd}}$, which is nontrivial in torsionful geometries, but conformal with certain homogeneity degree, doing the gravity persist even in the post-Newtonian limit when the torsion let of act, because begins exist the vacuum of the Universe.

But in the torsion case, the movement equations imply:

$$T^{\mu} = \frac{3\kappa}{4} \frac{\beta}{\beta^2 + 1} J_{(A)}^{\mu}, \tag{6}$$

$$S^{\mu} = \frac{3\kappa}{\beta^2 + 1} J_{(A)}^{\mu}, \tag{7}$$

$$q_{\mu\nu\rho} = 0, \tag{8}$$

where is clear an inconsistent due to the equality between a vector T^{μ}, with the pseudo-vector $J_{(A)}^{\mu}$. Then we conclude that gravity acts and gives shape and cohesion to the space-time after certain step, when the current term $\frac{\kappa}{4} \in_{\mathrm{cd}}^{\mathrm{ab}} e_{\mu}^{c} J_{(A)}^{\mu}$, is added (see **Figure 1**). The Immirzi parameter only is the structural parameter to solve the inconsistent in the equation systems (6)–(8), but no symmetry.

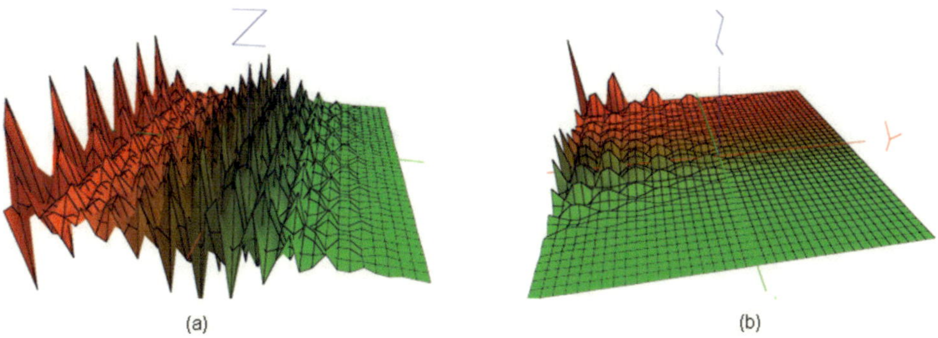

Figure 1. (a) Model of space-time with matter production without Immirzi parameter, but yes torsion. (b) Model of space-time with gravity and Immirzi parameter, including torsion. Observe the non-symmetric space created in the waving movement due to gravity.

The Holst term needs modification by the addition of fermions to become total derivative [3, 4], and provokes the coupling with the gravity to a global action on the space-time.

The fermion-piece that will be added to the Holst action is:

$$S_{\text{Holst-Fermion}} = \frac{\alpha}{2}\int \mathrm{d}^4 x e \times \left(\overline{\psi}\gamma^{\mu}\gamma_5 \mathcal{D}_{\mu}(\omega)\psi + \overline{\mathcal{D}_{\mu}(\omega)}\gamma^{\mu}\gamma_5\psi\right), \alpha = \text{cont}, \tag{9}$$

together with Dirac kinetic term, the fermion action reads:

$$S_{\text{Dirac-Holst-Fermion}} = \frac{\mathrm{i}}{2}\int \mathrm{d}^4 x e \times \left(\overline{\psi}\gamma^{\mu}(1-\mathrm{i}\alpha\gamma_5)\mathcal{D}_{\mu}(\omega)\psi + \overline{\mathcal{D}_{\mu}(\omega)}\gamma^{\mu}(1-\mathrm{i}\alpha\gamma_5\psi)\right), \tag{10}$$

If $\alpha = \beta$, the *Holst total gravity-fermion* term becomes total derivative, that is to say:

$$S_{\text{Holst}-\text{total}} = -\frac{\beta}{2}\int d^4 x\left[I_{\text{NY}} + \partial_{\mu} J^{\mu}_{(A)}\right], \tag{11}$$

which establish an equivalent to the Einstein-Cartan theory independent of Immirzi parameter β.

In Eq. (11), I_{NY} is the *Nieh-Yan topological invariant density* which creates the unique Lorentz invariant torsion structure, which has the form:

$$\mathrm{I}_{\text{NY}} \equiv \mathrm{e}^{\mu\nu\rho\sigma}\left(\mathrm{T}^{\mathrm{a}}_{\mu\nu}\mathrm{T}_{\rho\sigma\mathrm{a}} - \frac{1}{2}\mathrm{e}^{\mathrm{a}}_{\mu}\mathrm{e}^{\mathrm{b}}_{\nu}R_{\rho\sigma\mathrm{ab}}(\omega)\right), \tag{12}$$

which in simplified shape is written as

$$I_{\text{NY}} \equiv e^{\mu\nu\rho\sigma}\partial_{\mu}T_{\mu\nu\sigma}, \tag{13}$$

Then our model of cohesion by gravity is possible, since the scattering in the space-time is decreased under the last action (see **Figure 1(a)**).

3. Diffeomorphisms in gravity

Likewise, to energy level, let E be the energy, S be the space, and T be the time, and the corresponding topological spaces be U_S^E and U_T^E [5] such that the following commutativity of the diagram is satisfied:

$$\begin{array}{ccc} T & \xrightarrow{\xi} & S \\ f\downarrow & & \downarrow g, \\ E & \xrightarrow{\text{Id}} & E \end{array} \tag{14}$$

where in the context of the energy-(space-)time, an application ξ is an Universe homomorphism:

$$U_T^S = U_S^E U_T^E, \tag{15}$$

considering that $f: T \rightarrow E$ is a smooth mapping between manifolds (the composition $\xi \circ \psi$), and let U_T^E be the image of the universal covering established for the commutative diagram. Then the image U_T^E is a manifold of E (the space of energy). We say that f is transversal to U_T^E, which can be denoted as $f \cap U_T^E$, if and only if, to each $x \in f^{-1}\left(U_T^E\right)$, is had that:

$$\mathrm{Im}(\mathrm{d}f_x) + \mathrm{T}_{f(x)} U_T^E = \mathrm{T}_{f(x)} \mathrm{E}, \tag{16}$$

and analogy to the space S had a smooth mapping between manifolds S and E such as $g: S \rightarrow E$, where the image U_S^E is a submanifold of the space E. Then is had $g \cap U_S^E$, if and only if to $y = g^{-1}\left(U_S^E\right)$, is had that:

$$\mathrm{Im}\left(\mathrm{d}g_y\right) + \mathrm{T}_{g(y)} U_S^E = \mathrm{T}_{g(y)} \mathrm{E}, \tag{17}$$

Def. 3.1. Given that $f: T \rightarrow E$, and $g: S \rightarrow E$, the applications of f, and g, are transversals (that is to say $f \cap g$) and if $\forall x \in T$, and $y \in S$, satisfy $f(x) = g(y) =: z$, then is had that:

$$\mathrm{Im}(\mathrm{d}f_x) + \mathrm{Im}\left(\mathrm{d}g_y\right) = \mathrm{T}_z \mathrm{E}, \tag{18}$$

Then a diffeomorphism between the space $T_{g(y)}E$ and $T_{f(x)}E$, can be established, where f and g are differentiable (for their definition as mapping). By transversality, we have the following mapping [6]:

$$\mathrm{T}_{f(x)} E \rightarrow T_{g(y)} \mathrm{E}, \tag{19}$$

that establishes the variation of the energy-(space-)time states in the covering space E [4]. Their differential is the curvature. Their integral is the variation principle $\delta(\int Diff\, f)$.

But how the energy states vary in this broth of space-(energy-)time given by the Universe U_T^S, of such lucky that the matter can be formed with the step of time. Here, the time happens as distance between cause and effect or distance between initial state and final state of a waving in the space.

Let X, be a field acting in the space-time as energy that involves space and time [7]. An oriented string τ, in this space U_S^T, can be defined from a $\varphi_0 \in \mathrm{E}$, until a $\varphi_\alpha \in \mathrm{E}$, as a continuous function $X : [\varphi_0, \varphi_\alpha] \to U_S^T$, such that $X(\varphi_0) = \varphi_0$, and $X(\varphi_\alpha) = \varphi_\alpha$, in certain evolution step α [2].

Def. 3.2. We consider the Lagrangian $L(\varphi, \dot{\varphi})$, for the energy state $\varphi \in E$[3]. Then, we define the perception or field conscience operator O_C, as the mapping [8]

$$\mathrm{O_C : TE \to TE^*}, \tag{20}$$

with the correspondence rule[4]

$$\varphi \mapsto \mathrm{O_c}(\varphi), \tag{21}$$

which is translated in the differential

$$\mathrm{d}\mathfrak{I}(\varphi)\mathrm{h} = \int_\Gamma \left(\frac{\partial \mathrm{L}}{\partial \varphi} - \frac{\mathrm{d}}{\mathrm{dt}} \frac{\partial \mathrm{L}}{\partial \dot{\varphi}} \right) \left(\varphi(\omega(\mathrm{s})), \dot{\varphi}(\omega(\mathrm{s}))\mathrm{h}(\omega(\mathrm{s}))\mathrm{d}\omega. \tag{22}$$

We observe that the diffeomorphism given in Eq. (22) is the field conscience operator (20) with topological definition in Eq. (6).

The diffeomorphism O_C is of gravity if their connection [5, 7, 9, 10] of gravitational field $h(x)$, (considering the restriction condition given in Eq. (4)) is[5]:

$$(h(x)) - \frac{1}{2} g(x)h)x(s) = \eth(x(s)), \tag{23}$$

Then, the gravity conscience satisfies the identity by (20)–(22) as[6]:

$$\int_H \left(T(\varphi)h(\varphi) - L\left(\varphi, \dot{\varphi}\right)\right) d\varphi = \int_\Omega \left(\left(h(x) - \frac{1}{2} g(x)h \right) \right) x(s) d(x(s)), \tag{24}$$

where we have used the property of O_C, given in [8]:

---

[2] Remember that in QFT, field and wave can be the same using the two-duality principles: field and particles are the same and also wave and particle. Then by equality transitivity, field and wave are the same. Likewise, in string theory a string is a field.

[3] φ is an energy state of a field X.

[4] φ, X. $\mathrm{O_c}(\varphi) = \mathrm{d}\mathfrak{I}(\varphi) = L\left(\mathrm{s}, \varphi, \dot{\varphi}\right)\mathrm{d}\varphi$.

[5] The appearing of a vector $v(x)$, is when we derive again the expression (23). Then we obtain a vector as appears in the integrant of the integral expression given in Eq. (22).

[6] The integral in the right member is realized over the space $\Omega . \subseteq \mathbb{R}^3 \times \mathrm{I_s}$.

$$\int_{H} O_C(\varphi(x))d\varphi = \int_{\Omega} \mathcal{O}\left(x - x'\right)x(s)d(x(s)), \tag{25}$$

where $\mathcal{H}$ is the energy states space defined explicitly[7]:

$$\mathcal{H} = \{\varphi(x) \in [m] | [m] \subset \mathrm{TE}^*\}, \quad [m] = \mathrm{T}{*}\mathrm{C}_{\mathrm{n,m}}, \tag{26}$$

The operator $\mathcal{O}(x - x')$, derives from the study of functional derivative in quantum mechanics [8, 11]. $T(X)$ is the energy-matter tensor due to the source of matter production.

Expression (24) proves the Conjecture 1.1, given in Section 1, since this demonstrates that the field is a matter-production source (the particle process does appear 2-spin particles expressed inside the energy integral) that is to say, the gravity is the performance of the space-time, and O_C, induces a curvature tensor due the gravitational energy perceived for this by this performance.

Def. 3.3. The space-time performance is their gravity intention in certain evolution step.

Likewise, the gravity intention $\mathfrak{I}_{O_c}$ is to link the objects through a connection or shape operator of the gravitational field having the perception of matter given for the operator O_C,[8] (that is to say, for the only presence of the matter).

Finally, we can affirm considering the Dirac field equation and coordinates systems given for μ, ν, and λ that the diffeomorphism of gravity takes the invariant form:

$$\partial^{\lambda} G_{\mu\nu\lambda}(x) - \partial_{\nu} G^{\lambda}_{\mu\lambda}(x) + \mathrm{m}^2\left(\varphi_{\mu\nu}(x) + \frac{1}{2} g_{\mu\nu}\varphi(x)\right) = T_{\mu\nu} - \frac{1}{2} g_{\mu\nu} T(x), \tag{27}$$

Here is where appears the expression of difference of the energy-matter tensor to the equation of field (23), to the curvature tensor $R^{\mu\nu}$. Likewise, if the gravitational field persists as consequence of quantum interaction field in the Universe, then 0-spin particles and ½-spin particles must be transformed into 2-spin particles [12]. The conjecture 1.1 is totally demonstrated.

In [13], it was concluded that the torsion field stays univocally determined through $H-$ fields considering the established ranges (to matter and inflation [12, 13]). Also, the Dark matter begins their action on the fermionic dispersion to produce gravity annulling the chirality of the neutrinos/anti-neutrinos interaction and becoming these in oscillations of gravitons or "gravitational waves" registered and measured through spinor frames of the Majorana states.

Then this treatment to obtain a detailed description through microscopic context in field theory of the gravity (as matter-energy that is produced in the nucleon-synthesis, where is originated the matter and with it the fragmented Universe) is known with precision.

---

[7] Also we can to define this space considering the images to n- particles in m- systems (always with $n \leq m$) in $\mathrm{T} * \mathrm{C}_{\mathrm{n,m}}$, as: $\mathcal{H} = \{\phi(\mathrm{x}) | \mathrm{H}(\phi(\mathrm{x})) = \frac{1}{2\mathrm{m}} \|\phi\|^2 + \mathrm{V}((\phi(\mathrm{x}))\}$.

[8] Conscience of the presence of matter. This conscience is the only presence of matter that does to change the spin of energy of the space between material bodies. Remember, the energy change and provokes movement.

But, these material products (such as planets, stars, galaxies, nebulas, etc.) are followed and joined by wave-links of gravity which conform their orbits and relative positions in the Universe.

We could say that the dark matter is the result of energy-matter with hadrons in action.

4. Observational confirmations and conclusions

As was mentioned in the beginning of this chapter, the inflation is normed by the matter production and for the existence of inflatons (condensation of gravitinos, which come of the interactions in the leptogenesis (**Figure 2**)). This determines the first inflation in the double inflation determined in the conformal SUGRA model which corresponds to our string model τ of the space-(energy-)time given for our topological space U_T^S in TFT. Then the second inflation happens in the Starobinsky inflationary phase ($\Lambda \neq 0$) [15] due to massive gravitino quantum fluctuations (induced by the higher curvature terms in 1-loop effective action). In this phase, the baryon-genesis ready to produce 2-spin particles happens, that is to say, gravity. This is the phase where the shape of the Universe with curvature has place. The cohesion is realized from the conformal SUGRA model (see the Planck + WP + BAO area **Figure 3** [9, 13, 14]). The second inflation happens as consequence of the quantum fluctuation in the 2-spin particles waving [13]. The operator $O(x - x')$, through their functional derivative that defines, create the diffeomorphism defined by O_C, whose derivative is their Lagrangian. The integral given in Eq. (25) establishes the Huygens principle on second dynamics law in terms of energy, because for one side of Eq. (25), we have gravitational energy and, in the second member, we have a force of gravitational field called by Huygens in the century XVII as "live force" of the Universe.

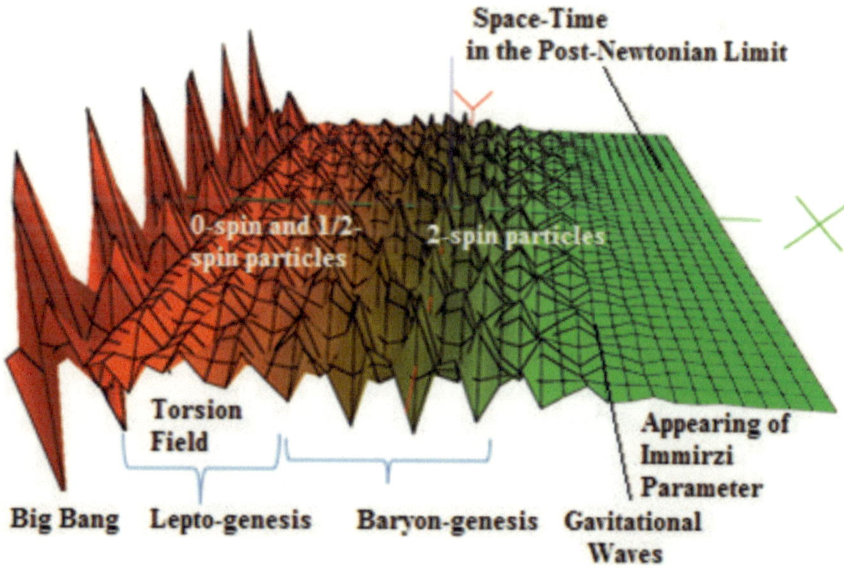

Figure 2. Evolution of the Universe and gravitational sequence of the energy-matter tensor [9, 12–14].

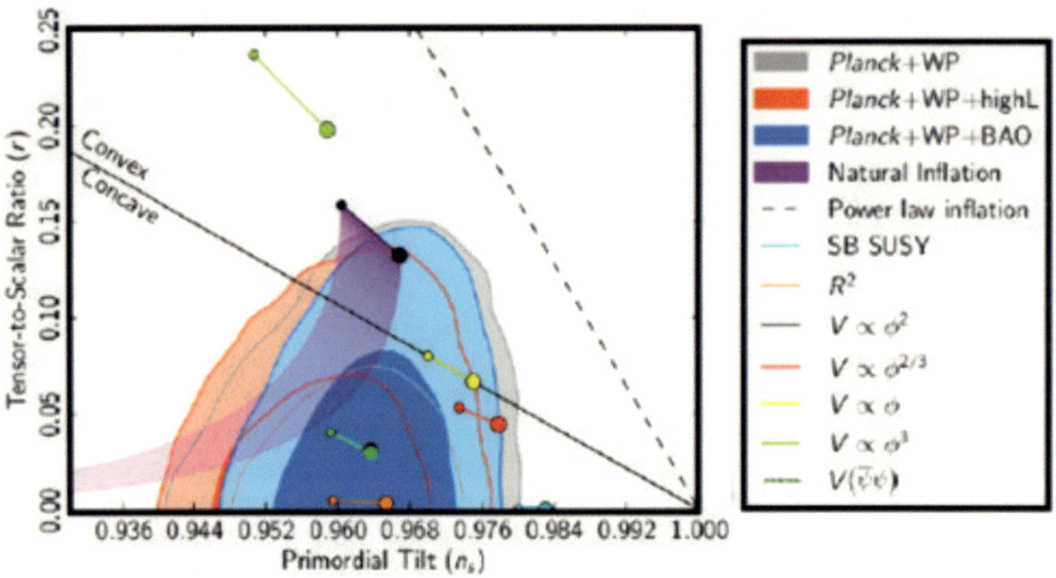

Figure 3. Energy-matter tensor development in the lepto-genesis and baryon-genesis processes. The SB SUSY region or area is where takes form the space U_T^S. the convexity in the thermal and expansion arrow meets to the causal structure of the universe defined by the light cones and given by the homomorphism (15).

Author details

Francisco Bulnes[1,2*]

*Address all correspondence to: francisco.bulnes@tesch.edu.mx

1 Research Department in Mathematics and Engineering, TESCHA, Chalco, State of Mexico, Mexico

2 International Advanced Research in Mathematics and Engineering (IINAMEI), Chalco, State of Mexico, Mexico

References

[1] Bulnes F. In: Pahlavani MR, editor. Correction, Alignment, Restoration And Re-Composition of Quantum Mechanical Fields of Particles by Path Integrals and Their Applications, Theo- retical Concepts of Quantum Mechanics. Janeza, Croatia: InTech; 2012. DOI: 10.5772/32847 Available from: https://www.intechopen.com/books/theoretical-concepts-of-quantum-mechanics/correction-alignment-restoration-and-re-composition-of-fields-of-particles-by-path-integrals-and-the

[2] Stropovsvky Y. Local diffeomorphisms and smooth embeddings to complex integral geometry, Chapter 2. In: Bulnes F, editor. Post-Doctoral Seminar on Curvature. EAE; 2017

[3] Bulnes F. Detection and measurement of quantum gravity by a curvature energy sensor: H-states of curvature energy, Chapter 7. In: Uzunov DI, editor. Recent Studies in Perturbation Theory. Intech; 2017

[4] Corradini O, Koyama K, Tasinato G. Induced gravity on intersecting brane worlds: Maximally symmetric solutions. Physical Review D. 2008;**77**:084006

[5] Bulnes F. Integral Geometry Methods in the Geometrical Langlands Program. USA: SCIRP; 2016

[6] Ramírez MA, Verkelov I. Transversality, deforming and curvature of space-time through axions, Chapter 4. In: Bulnes F, editor. Post-Doctoral Seminar on Curvature. EAE; 2017

[7] Francisco Bulnes, Topological Quantum Diffeomorphisms in the Field Theory and the Spectrum of the Space-Time, Proceedings of ICATMP 2017: 19th International Conference on Advances in Theoretical and Mathematical Physics. United Kindom; May 2017 urn: DOI:10.1999/1307-6892/10006924

[8] Bulnes F. Quantum intentionality and determination of realities in the space-time through path integrals and their integral transforms. In: Bracken P, editor. Advances in Quantum Mechanics. InTech; 2013. DOI: 10.5772/53439 Available from: https://www.intechopen.com/books/advances-in-quantum-mechanics/quantum-intentionality-and-determination-25 of-realities-in-the-space-time-through-path-integrals-and-t

[9] Mielke EW. Einsteinian gravity from BRST quantization of a topological action. Physical Review D. 2008;**77**:084020

[10] Ramírez M, Ramírez L, Ramírez O, Bulnes F. Coverings and axions: Topological characterizing of the energy coverings in space-time. Pure and Applied Mathematics Journal. Special Issue. 2014;**3**(6–2):6-11. DOI: 10.11648/j.pamj.s.2014030602.12

[11] Schwinger J. Particles, Sources, and Fields. In: Advanced Book Program. Vol 1. USA: Perseus Books; 1998

[12] Guivenchy E. Torsion Matter and their Corresponding Dirac Equations on Neutrinos and Matter/anti-Matter Asymmetries (Part I: scattering difference between Neutrinos/anti-Neutrinos), Journal on Photonics and Spíntronics, 2014;**3**(4):9-18

[13] Gorbuix S, Guivenchy E. Torsion matter and their corresponding Dirac equations on neutrinos and matter/anti-matter asymmetries (Part II: Gravitation insights). Journal on Photonics and Spintronics;**4**(2):15-25

[14] Shaposhnikov ME, Farrar GR. Baryon asymmetry of the universe in the minimal standard model. Physical Review Letters. 1993;**70**(19):2833-2836

[15] Kahya EO, Woodard RP. Scalar field equations from quantum gravity during inflation. Physical Review D;**77**:084012